ARCHAEOLOGY OF COASTAL CHANGES

PROCEEDINGS
OF THE FIRST INTERNATIONAL SYMPOSIUM
"CITIES ON THE SEA–PAST AND PRESENT"
Haifa, Israel, September 22–29, 1986

Edited by
AVNER RABAN
University of Haifa

Center for Maritime Studies, University of Haifa, Israel
In collaboration with C.A.H.E.P.,
Caesarea Ancient Harbours Excavation Project

Publication No. 2

BAR International Series 404
1988

B.A.R.

5, Centremead, Osney Mead, Oxford OX2 0DQ, England.

BAR S404, 1988 : 'Archaeology of Coastal Changes'.

ISBN 9780860545194 paperback
ISBN 9781407346847 e-book
DOI https://doi.org/10.30861/9780860545194
A catalogue record for this book is available from the British Library
This book is available at www.barpublishing.com

LIST OF CONTRIBUTORS AND THEIR ADDRESSES

Beatrice Basile, Giovanni Di Stefano, Gioacchino Lena
c/o Soprintendenza ai Beni Culturali e Ambientali, Piazza Duomo 14, 96100 Siracusa, Italia

Anne-Marie Collombier
7 Rue de la Chaine, 31000 Toulouse, France

Frank J. Frost
Dept. of History, University of California, Santa Barbara, CA 93106, U.S.A.

Gerd Hoffmann, Horst D. Schulz
Geologisch Institut der Universität Kiel, Olshausenstrasse 40, 2300 Kiel, Germany

Evangelos Kambouroglou
Ministry of Culture, 12 Karitsi Square, Athens, Greece

John Loret
Science Museum of Long Island, 1526 North Plandome Road, Manhasset, New York 11030, U.S.A.

Hampik Maroukian
Dept. of Geography, University of Athens, Panepistimiopolis, 157 84 Zografou, Greece

Martine Moerman
54, Avenue de St. Menet, 13011 Marseille, France

William M. Murray
Dept. of History, University of South Florida, Tampa, FL 33620, U.S.A.

Tina M. Niemi, Eberhard A.W. Finke
Dept. of Geology, Stanford University, Stanford, CA 94305, U.S.A.

Marinella Pasquinucci
Dipartimento di Scienze Storiche del Mondo Antico, Università di Pisa, Via Galvani 1, 56100 Pisa, Italia

Paolo Antonio Pirazzoli
CNRS — Intergéo, 191, Rue St. Jacques, 75005 Paris, France

Avner Raban
Center for Maritime Studies, University of Haifa, Haifa 31999, Israel

Gabriella Rossetti
Dipartimento di Medievistica, Università di Pisa, Italia

Bro. S. Dominic Ruegg
F.S.C., Saint Mary's College, Moraga, CA 94575, U.S.A.

Adamandios Sampson
Archaeological Museum of Khalkis, Greece

Serge Ximenes
Président de la Commission Nationale d'Archéologie Sous-Marine, F.F.E.S.S.M., 72, Traverse des Ecoles, 13011 Marseille, France

Graphic design: Zaraza Friedman

This volume is published with the generous assistance of The Caesarea Edmond Benjamin de Rothschild Foundation.

ACKNOWLEDGEMENTS

The editor of this volume is greatly indebted to the persons and institutions that helped in planning, organizing, hosting and handling the symposium and the proceedings. We wish to thank Prof. R.L. Hohlfelder, Dr. N.C. Flemming, and Dr. P.A. Pirazzoli for their constant encouragement in holding the international meeting in Haifa and for their active roles in promoting our case throughout the international community of scientists. Thanks are also due to Dr. Elisha Linder of the Center for Maritime Studies at the University of Haifa for being not only the original dreamer and "godfather" of the symposium and its special charter but also for spreading the gospel so successfully and for presiding over its scientific planning committee.

Professor Yehuda Karmon, Professor Emeritus at The Hebrew University of Jerusalem, the first Chairman of the Center for Maritime Studies and the President of the Symposium was the first Israeli academic to promote a coherent approach to the interdisciplinary study of harbours.

We are also indebted to our colleagues and co-organizers, Dr. D. Rosen and Prof. M.L. Vajda of the Technion's CAMERI, Dr. Y. Tzamir and Prof. M. Kashtan of the Faculty of Architecture and Town Planning of the Technion, Dr. A. Golik and Prof. V. Goldsmith of IOLR, and Dr. Y. Hayuth of the Department of Geography and the Wydra Institute of Shipping and Aviation Research at the University of Haifa. Thanks are also due to all the host institutions and all those in Haifa and elsewhere which provided assistance, especially to Prof. Colette Serruya, the Director of IOLR; Mr. S. Raziel, Director-General of the Ports Authority; Prof. A. Solan, Vice-President for Research at the Technion; Mr. A. Gurel, Mayor of Haifa; and Miss Trude Kerner of Unitours.

Mr. Y. Recanati, the general director of El-Yam Ships Ltd. has given us long-lasting backing and support. Mrs. Sara Baruchin and Admiral M. Limon, on behalf of the Baron Edmond de Rothschild, and the Caesarea Foundation have provided constant moral and financial assistance. Mr. N. Eshel, the Director of the Caesarea Development Company is to be thanked for acting as a wonderful host during the symposium.

This volume could not have been edited and published without the tedious work of its real editor and proof reader, Mrs. Nira Karmon, aided by other devoted members of the Center for Maritime Studies, Mrs. Shula Livne, Mr. Yitzhak Dagan, and Miss Zaraza Friedman. We owe special thanks to Professor Stanley Waterman for his work in re-editing and proof-reading most of the papers and for his most valuable suggestions for linguistic corrections.

Finally, we would like to dedicate this volume in homage to a great scientist and dear colleague whose ill-health made it impossible for him to attend the symposium that he had cherished so dearly. This volume is our homage to the memory of the late Professor Walter S.

Newman of the Department of Geology at Queens College of the City University of New York. May this volume be a humble additional contribution to his humanistic spirit of collegial interdisciplinary work and to his belief in a better understanding of the past for a better future for mankind.

A.R.

TABLE OF CONTENTS

INTRODUCTION

Archaeology of the Old World has always been a humanistic discipline with an accepted charter: to provide museums with artefacts and to illustrate the historical narrative. When marine archaeology was introduced as a subdiscipline, the general notion that pertained was that it would open many new opportunities. With the outstanding cases of the "time capsules" of Pompei and the Royal tomb of Tutankhamun in mind, many people envisaged the lost treasures of Atlantis and the wooden chest of Captain Kidd as representing two of many potential "time capsules" of the same kind. It is no wonder that marine archaeology is still considered by many, including scientists at some world famous academic institutions, to equate with nautical archaeology or underwater archaeology. Although most of the wrecks that have been excavated and studied so far have been found to be less exciting than the Etocha or the one at Kash, people still consider marine archaeology's real potential to be derived from the fact that a vessel that had sunk in water that was sufficiently deep might have remained unsullied by the elements and untouched by greedy human hands, and therefore a genuine "sunken treasure".

This prevailing concept which fits the classical attitude toward archaeology in general contradicts the present trend toward New Archaeology in land excavations. The New Archaeology was developed rapidly in countries outside the Biblical and Classical worlds. In northwestern Europe and in North America, written history might only illustrate the most recent stages of human cultures. For this reason, the study of earlier human societies in these regions cannot be guided by literal documents or even by the stylistic forms of the artefacts. The classical concept of archaeology as the "dirty maid" of the liberal arts has had to be replaced by a more realistic and scientific approach. Carbon 14 dating of organic substances, pollen analyses of ancient vegetation, the sedimentological study of palaeotopographic depositional environments, or the measuring of isotope ratios in the consideration of palaeoclimates are all gradually taking the place of artistic styles and even of modelling ancient societies according to the best formulae of social anthropology and other social sciences, as the best way for a better and more comprehensive understanding of the unrecorded ancient human heritage.

The New Archaeology would therefore consider that its charter was not the study of "time capsules" but rather the constantly changing mutual interactions between man and nature, between a human society and its surroundings. This represents a dynamic approach, rather resembling that of the geological history of our planet or of the history of a civilization. The students of the earth sciences were among the first to realize that interdisciplinary research could convert archaeology from being a client to a benefactor.

As prehistorians learned long ago, statistical classification of stone-made utensils is insufficient for understanding the Stone Age. As a result, they now work in firm collaboration with sedimentologists, palaeontologists, zoologists, botanists, and even nuclear physicists. On the other hand, the data on human societies, their habitats, their sources of nutrition and the physical leftovers of their daily life has been found to be extremely valuable to those scholars studying climatic changes and geomorphological processes. In this respect, the coastline is apparently more important than any other area. Minute climatic changes may occur, unnoticed, in many human societies, but on the seashore they have a devastating impact. An eustatic change in the sea-level of one or two metres might have been caused by an almost unnoticed warming of the atmosphere by a few tenths of a degree. However, along the coastal plains of the virtually tideless Mediterranean Sea, it might bring about a catastrophe to harbours, salinas and even to agriculture. The shoreline, which results from the interaction of marine and terrestrial processes, is always in a very delicate equilibrium. Each small change in the rates of transportation of the load of riverine sediments or of wave- carried sands or in wave climate can affect this balance. The populations that are settled on the coast are thus more vulnerable to such changes than are populations inland, so much so that even their attempt to accommodate the topographic changes by taking active measures might trigger other, indirect, changes along their own shores.

This meeting place between sea and land, where changes are wrought with ease both horizontal and vertical planes, is very often of prime importance for the populations of the coast. Regression would bring about a lowering of the water table on the coastal plain and fields and wells would dry up. Rising seas would cause them to become saline. A fall in the supply of sand up the coast would lead to additional abrasion of the shoreline. Excessive sedimentation would silt up anchorages and leave harbour settlements landlocked. The rates of such changes are often rather fast and frequently beyond management, even with our modern technologies. All this is well known to students of the earth sciences and civil engineering. Archaeologists and historians are usually less aware of its significance.

The aim of the symposium which was held at the University of Haifa was therefore to bring together the historians, archaeologists, oceanographers, geomorphologists, geologists, civil engineers, town planners, environmentalists, and other scholars who deal in various aspects of the ever-growing demand for land uses along the coast.

The underlying idea for this attempt is that notwithstanding the great developments in science and technology, we are still far short of an understanding of the full impact of what is taking place along the coast in terms of human intervention in natural processes. How can we have a full understanding of this when scientists are still far from being sure of what are the current trends of many of these processes? The basic concept of the symposium is that people have had to confront the constant problem of changing coastlines since they first settled alongside them and began to utilize the sea as a source of livelihood and as the best means of transportation over long distances. It is only logical to deduce that these coastal - or, to use a better nomenclature, Maritime - Civilizations, might have gained some experience and even found ways of their own for preventing their coastlines from being destroyed by nature or by human actions. It is also logical to assume that the history of coastal populations that were so vulnerable to topographic changes might have been dictated by such needs.

The idea of the symposium was originally brought up by the participants of the international workshop that was organized by the directors of CAHEP (Caesarea Ancient Harbours Excavation Project) in June 1983 (see B.A.R. International Series 257). The Center for Maritime Studies at the University of Haifa took upon itself the task of organizing and promoting the symposium and it soon discovered that other institutions welcomed the initiative and were willing to join in the organization and hosting of the symposium.

The co-operating institutions were the Departments of Geography and Maritime Civilizations at the University of Haifa, the Israel Oceanographic and Limnological Research, the Faculties of Civil Engineering and of Architecture and Town Planning, and the Coastal and Marine Engineering Research Institute of the Technion-Israel Institute of Technology. The symposium was held in Haifa where the hosts were the institutions listed above and in Caesarea where the sponsor was the Baron E. de Rothschild Caesarea Foundation which has made the publication of this volume possible. UNESCO's International Geological Correlation Programme (Project 200: Sea Level Correlations and Applications), Israel's Ports Authority, and the Municipality of Haifa, all took the symposium under their auspices. Some 60 scientists from 14 different countries participated and presented papers at the symposium.

The present volume is not the complete proceedings of the symposium. Though the interdisciplinary approach of the meeting prevailed throughout the discussions, the panel sessions and the field trips, the general notion was that the published contributions might be divided into three thematic categories:

1. Maritime history
2. Harbour engineering and coastal processes
3. Archaeology and geomorphology.

Only the third group of contributions are included in this volume. The first group, Maritime History, will appear in the Mediterranean Historical Review, vol. 3, no. 1 (June 88)

As we noted earlier, we are not the first to follow the concept of New Archaeology and to adapt it to Marine Archaeology. The real pioneers in this field have been Schmiedt who has carried out an extensive coastal survey along the Tyrrhenian Coasts of Italy, looking for archaeologically dated evidence for historical changes in land-sea relations (Schmiedt, 1972); Professor M.R. Bloch, a chemist who collected historical, archaeological and geomorphological data in order to reconstruct ancient sea-levels as a principal factor in the history of salt consumption (Bloch, 1976); the well-known marine geologist and oceanographer, Professor N.C. Flemming who, since his first archaeological experience at the submerged Greek harbour at Apollonia, in Libya, has become the leading scholar promoting the case of interdisciplinary research of coastlines (see his latest summary with C.O. Webb, 1986).

In France, this approach is still largely the domain of geologists and geographers. Nevertheless, Dr. P.A. Pirazzoli has been very successful in his attempts to establish the case for studying sea-level changes on a worldwide scale and as an interdisciplinary issue through the ongoing UNESCO project IGCP 200 (1985).

The Anglo-German cooperation in surveying ancient harbours in Greece and Turkey has brought together the historian-archaeologist D. Blackman, the classical archaeologist J. Schafer and the civil engineer, P. Knoblauch into considering geomorphological study as a prominent part of their basically architectural study (Blackman, 1973). D.L. Inman, the coastal oceanographer from the Scripps Institution, while travelling around the world as a prominent authority on coastal processes and on the impact of coastal structures on shorelines, has become aware of the importance of archaeological data in providing a better understanding of these issues, and has even encouraged the inauguration of a marine archaeological division in his oceanographic institution (Inman, 1974).

The initiatives by Flemming and Inman led to the first international conference to combine marine archaeology with the study of Quaternary coastlines (Masters and Flemming, 1983). This approach was soon followed elsewhere, as in the case of the international conference at Aix-en-Provence on the subject of "Archaeological Data for Coastal Changes" (organized by R. Paskoff and P. Trousset). The conference in Aix was perhaps the first of its kind to be acknowledged by archaeological institutions (the regional CNRS and the University of Aix), and had been preceded in this respect by a special issue of the popular French magazine Histoire et Archéologie - Les Dossiers (February 1981) which carried the subtitle "Sea-level changes and seismics in antiquity".

As a student of the discipline of archaeology and as a devoted admirer of the silent maritime civilizations of the Ancient World, I sincerely believe that the interdisciplinary approach of the Haifa Symposium will enable a real breakthrough not only in the understanding of the past but also in better coastal management in the future.

Avner Raban

REFERENCES

Blackman, D., 1973. "Evidence of Sea-Level Change in Ancient Harbours and Coastal Installations". In: D. Blackman (ed.), Marine Archaeology. (Bristol: Colston Papers, No. 23).

Bloch, M.R., 1976. "Salt in Human History", Interdisciplinary Science Reviews, I.4, 336-352.

Flemming, N.C. & C.O. Webb, 1986. "Tectonic and Eustatic Coastal Changes During the Last 10,000 Years Derived From Archaeological Data", Z. Geomorph. N.F., Suppl.-Bd. 62, Berlin-Stuttgart, 1-29.

Inman, D.L., 1974. "Ancient and Modern Harbors: a Repeating Phylogeny", Proceedings of the 14th International Conference of Coastal Engineering, Society of Civil Engineers, Vol. 3, American 2049-2067.

Masters, P.M. & N.C. Flemming (eds.), 1983. Quaternary Coastlines and Marine Archaeology. (Academic Press).

Pirazzoli, P.A., 1985. "Sea-Level Changes", Nature and Resources. UNESCO, Vol. XXI.4, 2-9.

Raban, A., 1985. Harbour Archaeology, B.A.R. International Series 257. (Oxford).

Schmiedt, G., 1972. Il Livello Antico del Mar Tirreno. (Firenze).

BEATRICE BASILE, GIOVANNI DI STEFANO, GIOACCHINO LENA

Landings, Ports, Coastal Settlements and Coastlines in Southeastern Sicily from Prehistory to Late Antiquity

ABSTRACT

The purpose of this article is to describe the basic outlines of some integrated research aiming at the elaboration of an archaeological chart of the landings, the ports and, in general, the coastal settlements or all those human activities connected with the exploitation of the sea. The research for this article concerns Southeastern Sicily, in particular the coastline which includes the provinces of Syracuse and Ragusa. The data presented, and which in the final compilation will be set up by topographical indexes from north to south and synchronically and diachronically, are derived from a comparative analysis of ancient sources and archaeological findings in relation to the studies of the geomorphological transformations that took place over the centuries in the stretch of coast considered, starting with the prehistoric era up to the period of late antiquity.

GEOMORPHOLOGICAL CONSIDERATIONS

The long stretch of coast studied, permits settlement phases of varying length to be identified. The settlements influenced by cultural or geomorphological factors, were suited to populations dedicated to maritime commercial trading.

Three areas differentiated according to morphological characteristics, can be recognized. These are called respectively: (a) the prevalently rocky coast area, (b) the swamp area and (c) the dunes area.

(a) The Rocky Coast Area. In this area, the coast is high in some places and low in others. It has intervals of sandy crescents which are more frequent towards the south, and which are prevalent in the entire northern sector. The coastline presents, characteristically, a series of inlets. Occasionally, where the coast meets outlets of the larger water courses, especially those having greater erosive capacity, the sea penetrates for short distances inland, forming a "ria" type of submerged coast. This active transgression is also confirmed by the erosion and corrosion forms created in the shelly limestone, which covers the sediments of the Miocene and Pliocene cycles in carbonate and marly facies.

Recent studies reveal that this transgression is due only to a very small degree to eustatic variations of the sea-level while the prevailing factor is tectonic submergence. Our data do not allow us to specify the extent nor the date except in some points. However, it appears that it is of Holocene age. In fact, deposits of the Tyrrhenian cycle, containing, in the intermediate level, Strombus bubonius Lamarck, have been found along the coast north of Augusta. Their levels, at Gisira, near Brucoli are found, respectively, between 8.5-2 m below sea-level; 1.5-0 m below sea-level; -3/-10 m below sea level (Di Grande Scamarda 1973) and testify after the depositional and emergence phase, to a lowering phase. The relationship between these and other contemporary deposits, noted in nearby areas, demonstrates that submergence is differentiated in time and space and has involved this entire stretch of coastline to some degree. An analogous situation is apparent in the Thapsos area where these deposits are substituted higher up by an abrasion shelf, in the middle of an esplanade with a thin, reddish layer of sand and pebbles, while a shelf can be seen down to a depth of 10 m (Kapitan 1970).

The Neolithic village of Stentinello, almost cut in two lengthwise by wave movement causing crumbling of a cliff which today is no higher than two m, indicates the proximity of the sea. However, as there is no direct contact with the sea today the settlement area was once more extensive than it is today. The same situation of erosion and crumbling is met south of Syracuse, in the entrenched village of Ognina. The two villages of Gisira and Massoliveri, located on terraces at a higher elevation and relatively far from the sea show barely any signs of submergence.

Further evidence indicating shoreline movement during the Bronze Age comes from the sites of the larger settlements of Thapsos and Ognina, both of which are situated on small elongated peninsulas which extend outwards from the coast. Thapsos is linked to the coast by a sand dune and Ognina by a rocky isthmus.

Several artificial grotto tombs dug into the cliff on the eastern and northeastern slopes of Thapsos are found separating the first from the second succession of terraced surfaces. When the sea is rough they are flooded and this situation was already noticed during the last century (Orsi 1899). This is thus partly independent of the eustatic factor. The position of these tombs does not permit the definition of the actual position of the coastline during the second millenium B.C.E.; it must have been, in any case, further back than the present one. This situation is confirmed by the greater extension of the sand dune on which the port area was built and which has recently been brought to light (Voza 1973). The islet of Ognina used to form the furthest extension of a long and narrow fluvial valley which gradually submerged the Holocene period. The isthmus uniting the island with the mainland must have been extant at least until the Byzantine era when a small basilica was built on its summit. Between the 16th to the 13th centuries B.C.E., it was undoubtedly longer, permitting the establishment of a permanent Maltese settlement (Brea 1964) at the mouth of the channel port. A Siculan tomb submerged to a depth of about 1 metre permits the measurement of a submersion rate greater than that of Thapsos and which according to Kapitan would have had values of 3.5 to 4 m.

For the Ancient Greek era, the data at our disposal are the channel-port of Brucoli in the north, and in the south, the Syracusan promontory of the Plemmyrion. In Brucoli, several grottos have been dug artificially at the cliff's base, inside the channel-port. These lead us to assume that work was undertaken on the river-bed. Even without knowing the date of the excavation of all the grottos (a continuous excavation from the Neolithic to the Renaissance, (Schmiedt 1972) we can, in any case, presume that during the Ancient era it did not extend 85 metres towards the mainland, as is the case today, nor was there excessive infilling of the river-bed, a feature which makes it unusable today.

On the Plemmyrion promontory, near Syracuse, the extraction of the yellow calcarenite known locally as "sesame seed" and used for archaic monuments, reveals the existence of a landing, outside the city. At the same time this leads us to consider that the coast had a greater extension. This is confirmed by the finding of submerged caves down to a depth of 1 metre (Lena and Basile 1986). The other Syracusan ports, above all Lakkios, further confirm coastal submergence. The study of Lakkios in particular, the location of which has recently been defined and delineated at a depth of 3 metres below current sea-level, attests to the present-day reduction of the exposed area, a hypothesis which has already been formulated by Cavallari (Cavallari 1884). The greatest extension of the island of Ortygia is also well demonstrated by topographical considerations concerning the linkage between the island and the mainland (Gargallo 1970; Kapitan 1986; Voza 1973).

For the Roman era, a different shoreline position is attested to by the continuation of the production and distribution of lime in the industrial area, and by the development of Plemmyrion during this phase. The base of the more external furnaces is found at a depth of 4 metres below the present sea-level and testifies, in this point, to a considerable submergence (Lena and Basile 1986). Other furnaces are found south of this spot. Their different position in regard to the sea-level indicates a noticeable withdrawal of the coast.

In this discussion of morphology we do not know how to deal with the Eloro area which shows partial traces of submergence (the latomies - ancient stone quarries - outside the city are partially submerged) but the dominant consequences of which are noticeable erosion of the promontories (part of the Greek city has crumbled into the sea because of the erosion of the turbidite cliff) and the erosion of the beach north of the city with the filling from the beach to the south where the port was located.

(b) The Swamp Area. South of Eloro, from Vendicari-Cittadella dei Maccari down to the entire southern triangle of Sicily, the sector indicated as the swamp area begins.

The present-day morphology, made up of a series of coastal ponds elongated in a direction parallel to the coast and cut off from the sea by sandy stretches, follows similar repetitive patterns starting from the Tyrrhenian zone. The sediments of this era that show up extensively near the sea, lead us to suppose that, when they were deposited, the coast had many inlets with a number of islets nearby. During the sedimentation, sandy mounds were formed, many of which are on the sea-bed, without being exposed at the surface and thus creating barriers (Colacicchi 1963).

During the evolution of the Holocene era the above were renewed in morphological terms. The development can be assumed as follows:
- formations of coastal ponds in the shelly limestone hollows, favoured by the presence of a thin clayey cover;
- formations of ponds on the extreme margins of the rias (such is the case with the pond of Mandra Nuova where, on Tyrrhenian calcarenite, greenish clays from 3 to 4 metres thick with lagoon fauna are spread);
- formation of coastal ponds which act as barriers due to the movement of the sands. At times, to keep the pond waters at an acceptable level drainage channels were dug into them;
- formation of coastal sand bars at the outer margins of the bays.

There are few indications along this stretch of coast of port areas in ancient times except for the Byzantine era sites at the two extreme ends of the Vendicari Swamp and the Longarini Swamp. This does not preclude the fact that traces of a Neolithic village have been indicated at 2 metres below sea-level on the Vulpiglia Beach of Pachino (Kapitan 1970). During the Classical era, coastal traffic was connected with extractive activities and the processing of fish products. In fact, stone quarries have been recently studied that start from the inhabited area of Marzamemi (situated above the town) extend towards the sea for about 250 metres, until the quarry front is met at 1 metre below the present-day level (Lena and Basile 1986). Further south, the Morghella Swamp might have made up the saline (the calcarenite cordon that limits it towards the sea has been cut to regulate the return flow of the waters) used for salting the tunny in a workshop in use from the 4th century B.C.E. to the 4th century C.E. (Bacci 1982-83). The salting basins extend today on the shelf at an elevation of 2 metres, but continue along the present-day beach and go under sea-level in the direction of the island of Capo Passero (Bacci, personal communication). This indicates a transgression which, is however, of lesser intensity than that found in the northern sector.

Rounding Cape Passero, the Longarini Swamp encloses this swamp area and some of the characteristics of the succeeding area are introduced. Inside the swamp, during reclamation works, the relic of a Byzantine ship was discovered. It was dated to the 4th century C.E. and had sunk while entering the port (Throckmorton 1973). The wide inlet of the port of Ulysses must have continued towards the interior of a sheltered basin, after which is became isolated from the sea by means of a long sandy cordon. The port site corresponds with the station of Apolline, mentioned in the guidebooks, and extends intermittently between the port and present-day Punta Castellazzo (Di Stefano 1984).

(c) The Coastal Dunes Area. Beyond Apolline, the coast is straight and without ports. It is characterized by long sandy stretches which are limited inland by a series of sand dune cordons. These are laid out parallel to the present-day shoreline and normally towards the prevailing winds. In many places they stretch for about 1 km towards the interior.

On the basis of archaeological data there is no evidence to suppose coastal submersions. The normal phases of eustatic variations of the sea-level can be seen here while the evolution of the coast can be almost entirely attributed to the normal phases of erosion and sedimentation. For the most part, these are related to the varying solid transport of the water courses.

In this sense, the port at the mouth of the Irminio River seems typical, connected, perhaps, with the settlement of the Maestro Landing, which functioned as a port for a long time, from the Archaic to the Norman era. Research still being carried out makes up presume that it was located in a bend of the river, not far from the mouth.

Its disappearance can be attributed to an increase in the solid transport of the river caused by the deforestation of the basin upstream. This was followed by a hanging phase, which involved the formation of coastal sand dunes, with the infilling of the coastal inlet involving the formation of swampland.

Is there a similar situation of a channel port to be found at the mouth of a river at Camarina? Here a sea wall can be seen stretching towards the open sea in a south-westerly direction, starting from the present-day river mouth. It served to prevent the silting up of the port because of the transportation of sediments along the coast in a northerly direction. The infilling of the harbour took place following the demolition of the sea wall during the 16th century C.E.

The existence of a channel port obtained by digging the river-bed in the lower course of the river (Pelagatti 1985; Uggeri 1974) is shown by a series of quays at right angles to the present-day course, which is definitely not coincident with the ancient one. The system was kept in equilibrium by the existence of a lagoon inlet, the swamp of Cammarana further inland, which was accessible to boats of low draught (Blackman 1976-77).

The opposite is the case with Caucana which replaced the Camarina Landing during the period of late antiquity (Pelagatti 1973). Its location can be identified with the northern extremity of Punta Secca in which a submerged dug circular shelf can be seen, in aerial photographs, and the entrance of which is open to the northwest direction. The findings of the remains here include an ample haulage landing, towing bollards and a long line of submerged rocks which could confirm its use as a port (Blackman 1976-77; Di Stefano 1982-83).

REFERENCES

Bacci, G. 1982-83. Antico stabilimento per la pesca e la lavorazione del tonno presso Portopalo, Kokalos XXVIII-XXIX: 343-347.

Blackman, I.D. 1976-77. Bristol University Expedition to Camarina, Preliminary Report, Kokalos XXII-XXIII: 607-614.

Brea, B.L. 1966. Abitato neolitico e insediamento maltese dell'età del bronzo nell'isola di Ognina (Siracusa) ed i rapporti tra la Sicilia e Malta dal XVI al XII sec. A.C. Kokalos: XII: 40-69.

Cavallari, F.S., Holm, A. 1884. Topografia Archeologica di Siracusa.

Colacicchi, R. 1963. Geologia del territorio di Pachino (Sicilia Orientale). Geologia Romana II:343-404.

Columba, G.M. 1906. I Porti della Sicilia.

Crescenzi, I., Gaffarini, S. 1965. Tentativo di ricostruzione paleogeografica. La Sicilia attraverso il Neogene ed il Quaternario, Rivista Mineraria Siciliana VI, 32.

De Fiore, O. 1920. I fenomeni sismici della Sicilia e delle isole adiacenti, Atti Acc. Gioenia XII, Mem. VIII: 1-85.

Di Grande, A., Raimondo, W. 1983. Lineamenti geologici del territorio siracusano tra Palazzolo, Noto e Siracusa (Sicilia sud-orientale), Bollettino Società Geologica Italiana 102: 241-260.

Di Grande, A., Scamarda, G. 1973. Segnalazione di livelli a Strombus bubonius LAMARCK nei dintorni di Augusta (Siracusa), Atti Acc. Gioenia Sc. Nat. di Catania Serie VII vol. 5: 157-172.

Di Stefano, G. 1980-81. Ricerche sulla provincia di Ragusa (1977-1980), Kokalos XXVI-XXVII: 756-763.

Di Stefano, G. 1982-83. Appunti per la carta archeologica della regione camarinese in età romana, Kokalos XXVIII-XXIX: 331-340.

Di Stefano, G. 1982. Scavi e scoperte archeologiche nel basso Irminio, Itinerari del 54° Distretto Scolastico di Modica: 75-107.

Di Stefano, G. 1984. Apolline, ricerche archeologiche al Castellazzo della Marza, il Quindicinale 30.12.82.

Di Stefano, G. 1985-86. Kamarina, storia della città dopo la distruzione romana e prime ricerche archeologiche, Pagine dal Sud: 1-12.

Gargallo, P. 1962. Exploring Off the Coast of Sicily, Archaeology XV: 193-196.

Gargallo, P. 1970. Alcune note sull'antica sistemazione dei porti di Siracusa. Kokalos XVI: 199-208.

Kapitan, G. 1986. Sul Lakkios, porto piccolo di Siracusa del periodo greco, Arch. Stor. Sir. XIII: 167-180.

Kapitan, G. 1970. Perlustrazioni sottomarini sulla topografia originaria e la situazione portuale dell'abitato preistorico nell'isola di Ognina, Sicilia Archeologica II: 43-56.

Lazzarini, S., La Rosa, A., Cappellani, G. 1965. Dieci anni di esplorazioni paleontologiche ed archeologiche nel siracusano, Arch. Stor. Sir. XI: 138-146.

Lena, G., Basile, B. 1986. Coastal Geomorphology and Exploitation of Lithic Resources (Latomies and Limekilns) in the Territory of Siracusa in Ancient Time, Thalassa, in press.

Malatesta, G. 1985. Paleogeografia e Paleobiologia dell'Era Glaciale, N.I.S.

Martin R., Pelagatti, P., Vallet, G., Voza, G. 1983. Le città greche, La Sicilia Antica, I, 3.

Mazzanti, R., Parea, G.C. 1977. Erosione della "panchina" sui litorali di Livorno e Rosignano, Boll. Soc. Geol. It. 110: 317-372.

Orsi, P. 1899. Thapsos. Monumenti Antichi dei Lincei, I: 757-760.

Pelagatti, P. 1973. Camarina ed il territorio camarinese. Kokalos XIX-XXI: 354-357.

Pelagatti, P. 1985. Camarina. Bibliografia Topografica della Colonizzazione Greca in Italia e nelle Isole Tirreniche: 1-32.

Ruggieri, G. 1959. Geologia della zona costiera di Torre Vendicari (Sicilia sud-orientale), Riv. Min. Sic. X.

Schmiedt, G. 1964. Contribution of Photointerpretation on Reconstruction of the Geographic-Topographic Situation on the Ancient Ports in Italy, Atti X Congr. Int. Photointer., Lisbona: 16.

Schmiedt, G. 1970. Atlante Aerofotografico delle Sedi Umane in Italia, IGM, II: 125-126; I, tav. 75.

Schmiedt, G. 1972. Il Livello Antico del Mar Tirreno, Olschki: 121-148.

Tiné, S. 1961. Notizie preliminari su recenti scavi nel villaggio neolitico di Stentinello, Arch. Stor. Sir. VII: 1-13.

Throckmorton, P., J. 1973. The Roman Wreck at Pantano Longarini, I.J.N.A. 2.2: 243-266.

Uggeri, G. 1974. Edizione Archeologica della Carta d'Italia al 100.00 F. 175, Scoglitti, IGM I: 1-28.

Voza, G. 1973. Thapsos: resoconto delle campagne di scavo 1970-71, Atti XV Riun. Scient. Ist. It. Preist. Protost.: 133-145.

THE SYRACUSAN COASTAL ARC

Ever since the middle of the Neolithic period, there has been intensive frequenting of the Syracusan coast, with the systematic use of the zone connected with it. This coast has an indented shoreline, with many inlets and promontories and small sheltered bays, and offers favourable conditions for settlement. At present, the coastal arc is studded with a series of villages that possess a substantially similar topographical situation. Located in the immediate vicinity of the sea, they are made up of clusters of huts held up by poles planted into the rocky plain, invariably grouped around small bays or on low promontories that dominate sheltered landing-places, often protected by a deep moat or a series of discontinuous ditches. Among the many villages identified (Gisira, Megara Hyblaea, Stentinello, Cape Santa Panagia; Matrensa, Terrauzza, Cape Arenella, Ognina Vulpiglia), the best known is the settlement of Stentinello, with rectangular huts built on a low rocky terrace on the sea, which today has partly eroded the village area. A moat, reinforced by a stone-pile trench, circumscribes the village area. The repeated presence of defensive structures perhaps demonstrates a constant state of conflict. It undoubtedly also indicates that these villages constantly gravitated towards the sea. This was not exclusively for fishing and proves that the Sicilian Neolithic, was characterized by the establishment of a dense net of contacts with the world outside Sicily. These successive overseas links profoundly modified the structures of late Neolithic society. The birth of the first forms of commerce, the basic purpose of which was to furnish raw materials useful for the increasing needs of a more complex and technologically evolved society, is shown by the presence of stones (such as basalt and obsidian) not found in the local lithology.

The commercial nature of some of these Neolithic settlements lasts continuously through all of the metal ages and beyond. This is so especially in the case of Ognina, a site characterized by the presence of an islet, which was once united to the mainland by an isthmus which is submerged today. This isthmus protected the entrance to a channel port from the north. The port originated from an ancient "quarry" of fluvial origin, which is characterized even today by the presence of brackish water springs from under the sea floor. The most evident morphological feature of the site in ancient times, when the sea-level was certainly 4-5 m lower than that of today, was the presence of two excellent ports, one consisting of the submerged quarry, the other of the roadstead north of the isthmus. The second port was originally protected towards the east by a stone-pile dam the remains of which are submerged today. In this part of the coast which was so favourable to settlement, two Neolithic villages of Stentinellian type were set up (one on the coast north of the isthmus, the other on the present-day islet). These were drawn preferentially towards the little port to the north; later there was a settlement of Maltese origin, which lasted throughout all of the ancient and Middle Bronze Age and was characterized by the presence of a large quantity of Tarxien, Borg-in and Nadur style ceramics which demonstrate the presence of a fully-fledged market of Maltese people even more than prosperous commerce.

Apart from the absolutely specific nature of the Ognina settlement, an identification has been made. Within the basically agricultural Castelluccian facies some of the villages show a more particular "coastal" aspect. The topographical situation of these villages indicates functions that go beyond the simple control of the coastline in order to protect the routes against internal penetration, (a specific duty of the villages placed at the mouths of the quarries towards the sea), entering into the commercial sphere. The evident Aegean-Anatolian influences that characterize so many aspects of Castelluccian civilization, from the artisan products to the funeral and defence structures are proof of this. Among the category of villages projected toward the sea are some settlements of which there are only slight traces of housing structures. Among these are the village of Thapsos, located on the central plain of the Magnisi peninsula, of which some sections of the protecting wall with semicircular towers remain. The village corresponding to the necropolis of the Reale estate, was situated on the crag of the Epipoli tableland; the village of Gisira, on a promontory that dominates a small bay suitable for anchorage to the north and an excellent channel-port to the south, made up by the torrent of Porcheria (the ancient Pantakyas). This has been identified by many with the Greek site of Trotylon, a transitory settlement of the Megarese in an era when the colony was founded and much used as a settlement on a channel-port until the Middle Ages. The village of Calafarina on the Cugni was on a plain dominating a quarry, the external part of which even today constitutes an excellent harbour inlet.

The settlement of Thapsos during the Middle Bronze Age is better known. During this period the tiny peninsula, characterized by two excellent natural ports on the flanks of the narrow isthmus, saw the flowering of an exclusively maritime centre which was open to a prosperous overseas trade that took part in the Mycenaean and Maltese routes, as is testified by the ceramic materials and the influences on the housing from the 15th to the 12th centuries B.C.E. Other villages, characterized by similar favourable topographical features located near the sea, have been identified at various points along the Syracusan coast. Good examples are Cozzo del Monaco, located on the banks of the Molinello River which was navigable at that time, about 1 km from the river-mouth, and that of the Plemmyrion, on the southern shore of the port of Syracuse.

A less well marked emphasis on the sea, caused by specific historical events which created the basis for the birth of strong densely populated centres, became entrenched in the hinterland in defensive positions, characterizes the following period, which lasted until the beginning of Greek colonization. This period is characterized by the cultural facies of Pantalica and Cassibile. The ports of call functions in relation to a particular hinterland (Ortigia, the port of call of the Hyblaea-Pantalica area, Punta Castelluzzo, with a similar function with regard to the Xuthia-Lentini area) are known only, about certain centers (some of which are still under study).

For the Greek and Roman periods, Syracuse is the city for which the greatest number of data have been gathered concerning the port arrangement. The earliest settlement there was located around the present-day Little Port. At that time, the Greek Lakkios presented a considerably different arrangement due to the sea-level variations. Like the channel-port of Ognina, it was also made up of a submerged fluvial valley with a somewhat sinuous course near its mouth. It was flanked by stone walls, submerged today, but which at the time served as excellent breakwaters, offering shelter to shipping whatever the direction of the winds. Recent research, gathering data from fragments of structures which survived a series of poorly executed dredgings, have brought to light the existence of conspicuous dockworks on the northern shore of Lakkios. These were probably given a monumental façade by the addition of columns. Of the large bend of the present-day Big Port, only the more internal part was probably used (corresponding to the present-day Little Port); the arsenals mentioned by Thucydides (VII, 25) were located here and traces of these are still visible, made up of deep parallel channels dug into rock.

In the immediate vicinity of the city other areas have been located (particularly the bay of Santa Panagia, commonly identified with the Trogylos mentioned by historical sources and the bay of Massoliveri), probably used for maritime activities that required a less extensive use of structures and which was perhaps related to the outer suburbs. This latter is, without doubt, the case of the bay of Massoliveri, which makes up the most significant example of a specific aspect of the Syracusan area's economy during the Roman imperial era, i.e., the flourishing of a series of small maritime ports of call, grouped along the coast south of the city, tied to the production and trade of lime, the furnaces of which have been identified in significant numbers.

Our knowledge about the ports of the other major centres which grew along the Syracusan coast, during the Greek period is still uncertain and scanty. Very little is known concerning the ports of Megara, which probably used the mouth of the Cantera as a channel-port. In this case, a long construction of blocks forming a submerged shelf today is commonly thought to be, if not a pier, then the foundation of a building (a portico or some similar structure) connected with maritime activities. As regards ancient Eloro, a subcolony of Syracuse, it has only been possible to assume the location of the port at the mouth of the Tellaro River, on the basis of an aerial photograph. The course of this river from antiquity to the present time, has moved towards the north. Furthermore, it is worth remembering that the greater part of the promontory on which the city grew, today mostly eroded, assured the channel-port of good protection from the northerly winds.
For these centres, the port problems remain and new data gathering may derive from further digs and from geophysical soundings.

Finally, in the Late Roman and Byzantine periods, the coast shows the same breaking up that distinguished the different hinterland settlements; the greater part of the points favourable for landing - including the same ones that had been occupied in the earlier periods - are used for setting up population centres, mostly of modest size. Among these are the Byzantine hamlets of Ognina, Eloro and the larger Portopalo. But among these hamlets, some stand out for their greater density and for the traces of a more intense commercial activity, aimed above all towards the trading with Africa. Particularly interesting, and today the subject of research, is the settlement of Punta Castelluzzo, once the maritime landing of a vast territorial area in the Protohistorical Era, situated on a high promontory between two inlets. The one to the south, consisting of the mouth of the Castelluccio Torrent and perhaps originally used as a channel-port, does not contain archaeological traces, because of modern works which have straightened, with artificial banks, the mouth of the river. The wide bay to the north, recently explored by underwater prospecting, has yielded a conspicuous quantity of amphora materials of various origins and vase fragments that demonstrate its use as a landing both during all of the Late Roman era as well as the Byzantine Period, along with a considerable volume of trade. Finally, another significant settlement, and still poorly explored, is that of the anonymous centre located at Cittadella dei Maccari, on a high tongue of land looking out upon the sea and surrounded on two sides by large swamps. The more southerly of the two contains the preserved existence of the ancient port, identified, hypothetically with the φοινίκους λιμήν noted by Ptolemy (Geogr. III, 4,4) and connected with the sea by a channel out into the sand-dunes. The settlement, of which only some small religious buildings are known, seems to show, on the basis of recent studies, a greater consistency and also a greater antiquity (the first imperial era) than had been believed before; and its port was probably the maritime landing used by another settlement of the Late Roman era, recently discovered, located on a small hill (Cozzo, Gerbi) in the nearby hinterland.

In the whole arc of the Syracusan coast, these later dated villages, except for the chief centres, generally last until the Arab invasion and then only some, like Portopalo (the "Marsa al bawalis", that is "the port of the swamps" of Edrisi) survive.

REFERENCES

Agnello, S.L. 1953. Scoperta di una piccola catacomba a Portopalo (Pachino). Riv. Arch. Crist. XXIX: 167-183.

Bernabò Brea, L. 1966. Abitato neolitico ed insediamento maltese dell' eta del bronzo nell'isola di Ognina (Siracusa) e i rapporti fra la Sicilia e Malta dal XVI al XIII sec. A.C. Kokalos XII: 40-69.

Bernabò Brea, L. 1968. Il crepuscolo di re Hyblon. La Parola del Passato CXX: 161-186.

Fallico, A.M. 1969. Villaggi tardo-romani e bizantini della Sicilia Occidentale noti all'Orsi e loro attuale consistenza. Atti I Congr. Arch. Crist. Matera: 177-183.

Gargallo, P. 1970. Alcune note sull'antica sistemazione dei porti di Siracusa. Kokalos XVI: 199-208.

Kapitan, G. 1967-68. Sul Lakkios, porto piccolo di Siracusa del periodo greco. Arch. St. Sirac. XIII: 167-180.

Kapitan, G. 1970. Perlustrazioni sottomarine sulla topografia originaria e la situazione portuale nell'abitato preistorico nell'isola di Ognina. Sic. Arch. XI: 43-54.

Lazzarini, S., La Rosa, A. & Cappellani, G. 1965. Dieci anni di esplorazioni paleontologiche ed archeologiche nel siracusano. Arch. St. Sirac. XI: 139-143.

Mentesana, M. 1967. La Gisira. Notiziario Storico di Augusta 1: 7-15.

Orsi, P. 1889. Appunti per la paletnologia di Siracusa e del suo territorio. B.P.I. XV: 48-58.

Orsi, P. 1891. La necropoli Sicula del Plemmyrion. B.P.I. XVII: 115- 139.

Orsi, P. 1895. Scoperte paletnologiche nella Sicilia, B.P.I. XXI: 51.

Orsi, P. 1898. Noto. Avanzi di un centro abitato dei bassi tempi riconosciuto nella penisola detta Cittadella. Not. Sc.: 36-37.

Orsi, P. 1898. Necropoli sicule ai Cugni di Calafarina. B.P.I. XXIV: 163.

Orsi, P. 1910. Villaggio neolitico di Stentinello (Siracusa) B.P.I. XXXVI: 66-67.

Tiné, S. 1961. Notizie preliminari sui recenti scavi nel villagio di Stentinello. Arch. St. Sirac. VII: 113-116.

Tusa, S. 1983. La Sicilia nella Preistoria. Palermo.

Voza, G. 1972. Thapsos. Atti Riun. Scient. Ist. It. Preist. Prot. XIV: 192-204.

Voza, G. 1973. Thapsos. Atti Riun. Scient, Ist. It. Preist. Prot. XV: 133-145.

Voza. G. 1976-77. L'attività della Soprintendenza alle Antichità della Sicilia Orientale. Kokalos XXII-XXIII: 562-568.

THE RAGUSAN COASTAL ARC

Coastal villages of the Ancient Bronze Age have only been proven to exist in any numbers, along the coast of southeastern Sicily. These can be dated to between 1800 and 1400 B.C.E.

Our knowledge concerning other coastal sites of the Neolithic Era and the Middle Bronze Age is scarce and fragmentary. What there is seems to close the exceptional phenomenon of the quantity of coastal sites that flourished, during the Castelluccian facies of the first Bronze Age.

This singular concentration of prehistoric villages occurs, above all, in the area between Punta Secca and Kamarina (Orsi 1901; Orsi, 1903; G. Di Stefano, 1978). Here, small villages consisting of a few huts with a stone perimeter and roof-tops of branches and straw arose on the rocky crests and cliffs, directly behind the sea. Recent and past finds have enabled us to locate an authentic network of settlements.

The coastal areas chiefly indicated by these findings are Punta Zafaglione, Camarina, Passo Marinaro, Cozzo Campisi, Cozzo Ciaramiri, Branco Grande, Branco Piccolo, Punta Braccetto, Fimmina Morta (Orsi, 1901; Orsi, 1903; Uggeri, 1961; G. Di Stefano, 1978).

There are geographical, topographical, climatic and hydrographical factors that have favoured the phenomenon of settlement of these villages along this stretch of coast only (Di Stefano, 1978). These villages survived mainly on nautical activities, though there was also a mixed economy and diet, typically based on agriculture and pastoral activities and seafaring, as was always the case among the ancient populations of the Mediterranean Basin.

The rocks of Punta Secca, which jut into the sea, protected the coast from the strong eastern winds, which push the sea currents from the east even more in Antiquity than today. In this stretch of coast, beyond Punta Secca, because it is difficult for the currents to make the sea rough, it was easy to develop small craft which hugged the shoreline and which allowed fishing among the rocks. Furthermore, we have been able to observe that all of these coastal settlements are sheltered westwards by a low promontory of land, so as to form a small bay completely sheltered from the westward winds and currents.

This type of geological formation on this stretch of coast of southeastern Sicily, has favoured a curious and peculiar hydrographical characteristic made up of fresh water springs which flow directly into the waterline and even into the sea.

This phenomenon, observed on the beaches of Branco Grande, Corridore and Torre di Pietro, is due to the fact that the waters of the lowlands of Passolato, Cavette, San Martino, Fontana and Cannatello seep deeply into the river bed of impervious geological layers, forming real underground streams that eventually exit along the seashore. This created a very accessible water supply for the prehistoric coastal villages.

Even though the subcoastal belt in which these prehistorical archaeological site traces have been found have low annual precipitation, it is often freshened by moist southerly winds from the sea, creating pockets of humid air and leading to local atmospheric phenomena which render the immediate hinterland considerably more fertile. The most important of these prehistorical sites is that of Branco Grande excavated by Paolo Orsi in 1907 (Orsi, 1907; Orsi, 1912; G. Di Stefano, 1985).

The village, made up of about 40 oval huts, was built on a rough, rocky cliff directly over the sea. It is the only rocky stretch to be found along a vast, sandy and harbourless strand. The huts of this village were closely grouped together on the rocky flatland, in small clusters. The structure of these dwellings is well known by now. They had a stone perimeter on which perishable material must have lain. The hut floors are a mixture of gravel and clay. Seats, millstones, small fireplaces and pithoi have been found inside these huts. The utensils found are rich and quite varied (engravers, blades, scrapers), as is the ceramic equipment (cups, bowls, buckets). This village was also fortified by an effective defensive embankment - a massive wall about 3 metres high with a counterscarp. It was a fully-fledged defence wall probably built for the completely exposed position on the sea of this site.

It is likely that this site must have been seriously and continually threatened from the sea and foreign invaders so as to necessitate the building of such an imposing embankment. The route of this fortification follows the natural perimeter of the cliff on which the village was built. Within the spectrum of the prehistoric Sicilian settlements the Branco Grande site, being a fortified village, occupies a position of considerable interest, as it is practically a unique model. On this characteristic, this village resembles few others from the Eastern Mediterranean (at Siros and Syphnos), in the Central Mediterranean (at Borg in Nadur and Malta) and in the Western Mediterranean (at Lebous, at Los Millares at Pedra de Ouro and at Zambejal).

Near the village, among the rocks, long artificial excavations shaped like funnels have been discovered, where sailing craft could be pulled ashore. This accentuates the maritime character of this site which like other contemporary sites, set up along this stretch of the coast, can be considered among the few Sicilian prehistoric settlements dedicated to seafaring. These coastal villages, though situated along the sea, have never received direct cultural influences outside Sicily, as in the case of the Ognina site, but have felt only generic, cultural contributions that came from outside Sicily, especially of Maltese origin.

As concerns Classical Antiquity, the only large and important port structure of the eastern Sicilian coast, from Syracuse to Gela, is the port of the Greek city of Kamarina (Orsi, 1899; Orsi, 1903; Orsi, 1904; Pace, 1927; Pelagatti, 1962; Pelagatti, 1978). Kamarina has a Syracusan foundation. The third subcolony of Syracuse, it existed, with many ups and downs, from 598 to 258 B.C.E. (Pelagatti, 1978). The city-state of Kamarina, although established as a colony consisting of an agricultural population, immediately revealed strong commercial tendencies. The ancient history of the city, (destroyed by Syracuse in 553 B.C.E., (Pelagatti, 1979) a generation after its foundation, because of a political uprising against the mother-country during which it had as allies the Sikels of the hinterland and other Greeks), is conditioned precisely by economic and commercial reasons. The strategic position acquired by Kamarina for all of the trade to and from the hinterland and the equally strategic position it soon occupied in the centre of the Sicilian Channel, made it a shipping point for the abundant agricultural products of the hinterland and as the place of arrival for the thousands of archaic amphorae inside which were carried oil and wine (Pelagatti, 1980).

These amphorae, which are a measure of the flourishing Camarinese commercial activities are imported from Attica, Corinth, Chios, Lesbos, Etruria and North Africa (Pelagatti, 1980). The city was laid out with parallel streets according to an urban planning scheme of the Hippodameus type, with an Agora and a very long fortification which circumscribed an urban space already defined at the moment of foundation and which was gradually occupied by the city blocks. The town is, therefore, a city-state with a precise commercial aspect and a strong seafaring tendency.

The port which has been surveyed several times (Blackman, 1976; Pelagatti, 1976) led from the coast into the estuary of the Ippari River, behind which there was a big swamp. The port was, therefore, a really fully-fledged channel port with a breakwater starting from the promontory (Uggeri, 1974). Various port structures of the Classical Era are also known, such as buffers, piers and buildings (Uggeri, 1974; Pelagatti, 1976).

The port, which remained active until the Hellenistic Era, was continually filled with sand during the Middle Ages. As a result of recent research it is now known that east of Kamarina there is a site of an ancient Greek emporium located at the mouth of the Irminio River, in the Maestro district (Di Stefano, 1986).

This site used the mouth of the Irminio River as a port-shelter for the bartering of products. It was a flourishing market-place, and is still being studied. The dwellings during the Ancient Period occupied the first and highest hillock to the left of the mouth of the Irminio River, where various dwelling-places, made up of two or three rooms filled with transport amphorae placed along the sides of a big courtyard, appear to be divided up into large or walled-in spaces (G. Di Stefano, 1986). This anonymous emporium, without an autonomous territory of its own and bound to the turbulent vicissitudes of the Camarinese epoch of the Archaic Age, constituted a strategic point of arrival for various commercial flows, due to the mouth of the Irminio River which lay below, and which was adopted as a mooring-place.

The use of the river estuary as a channel-port continued at least until the beginning of the Middle Ages. Only afterwards did the mobile coastal sand dunes covered over with the Mediterranean bush of juniper and gum trees invade the sea passage of the channel-port of the Irminio River emporium (G. Di Stefano, 1986).

The Roman Empire, the Byzantine and early Mediaeval periods saw a considerable flourishing of ports, landings, anchorages and piers in the coastal stretch of southern Sicily. Mesopotamio, Torre di Pietro, Cymbe, Rosacambri, Kaukana, Maulli, Hereo, Donnalucata, Sampieri, Punta Religione, Pozzallo, Marza, Apolline, are known from itinerary sources and from documentary sources of archives and from topographical analysis (G. Di Stefano, 1982/2). Renewed trading interests, related to an intense commercial traffic or to a transit of strategic military nature directed towards the North-African and Maltese coasts, (G. Di Stefano, 1982/2) were concentrated along the coastal arc during late Antiquity.

The vitality of this stretch of the Sicilian coast during this period is reflected by a dense population in the hinterland into which artistic, architectural and cultural currents that originated outside Sicily, also filtered, certainly from North Africa and even from the Syrian- Anatolian region (G. Di Stefano, 1982/1 1982/2).

Two ports of call are dominant in this period: Kaukana and Apolline. Kaukana is a chorion from which Belisarius first sailed in 553 C.E., for the Greek-Gothic War and, afterwards, in 1091 C.E. Roger I.

The dwellings of Kaukana, scattered along the coast and around the port are known to have been fortified houses of more than one floor, and courtyards in a central position (Pelagatti, 1972). The site was lived in from the 4th to the 6th centuries C.E. The port was built, and artificially dug, inside the inlet of Cape Scaramia. Much wreckage of the Byzantine Era and the 11th century C.E. can be found sunk in the anchorage place.

The port of call of Apolline is situated in Punta Castellazzo (G. Di Stefano, 1982/3). It is the most southerly of the ports of Sicily, the last port of call for the ancient routes for Malta. Explorations at sea have uncovered the existence of port structures. The dwellings, which were, perhaps, a statio of the coast road between Agrigento and Syracuse, are being excavated on the little peninsula of Castellazzo. In this area, evidence has been found of a cult dedicated to Apollo Lybistinus, connected with an invasion by Libyans who were repulsed by a virulent pestilence unleashed by the god. But here there are also other localized cults connected with later discoveries of the passage of Ulysses in Sicily. Here, in fact, is located the port of Ulysses and the cenotaph of Hecata (G. Di Stefano, 1982/3).

The Arab place-names (Marsa al B. Walis; Marsa) and the monumental remains demonstrate that the structures were used until the 15th century C.E.

REFERENCES

Blackman, D. 1976. Bristol University Expedition to Camarina, Preliminary Report in Kokalos, XXII-XXIII: 607-614.

Di Stefano, G. 1978. Villaggi Castellucciani sulla costa di Camarina, in Magna Grecia, a.XIII nn. 3-4, 17-20.

Di Stefano, G. 1982/1. Influenze microasiatiche nell' architettura religiosa dell'entroterra di Caucana, in Cronorama, X, No. 28-29; 25-33.

Di Stefano, G. 1982/3. Appunti per la carta archeologica della regione camarinese in èta romana, in Kokalos, XXVIII-XXIX: 553-560.

Di Stefano, G. 1982/3. Apolline: ricerche archeologiche al Castellazzo della Marza, in Il Quindicinale 30.12.82.

Di Stefano, G. 1985. Piccola Guida alle Stazioni Preistoriche degli Iblei: 47-49.

Di Stefano, G. 1986. L'Emporio Greco dell'Irminio.

Orsi, P. 1899. Camarina, campagna archeologica del 1896, in M.A.L., IX: 10-86 (cc. 201-222).

Orsi, P. 1901. Tracce di officine neo ed eneolitiche presso Scoglitti, in B.P.I.: 273-281.

Orsi, P. 1904. Camarina, campagna archeologica del 1899-1903, in M.A.L., XIV: 6-200 (cc. 757-952).

Orsi, P. 1907. Villaggio siculo sulla spiaggia di Camarina (Siracusa), in B.P.I. a XXXIII, No. 1-5: 45-46.

Orsi, P. 1910. Due villaggi del primo periodo siculo, I. Il villaggio preistorico di Branco Grande presso Camarina, in B.P.I., a XXXVI: 159-193.

Pace, B. 1927. Camarina.

Pelagatti, P. 1962. Camarina, relazione preliminare della campagna di scavi 1961-62, in B. d'A.: 251-264.

Pelagatti, P. 1972. Camarina e il territorio camarinese, in Kokalos, XVIII-XIX: 182-184.

Pelagatti, P. 1976. Nuove ricerche lungo la costa di Camarina, in Sicilia Archeologica, IX, 30: 15-28.

Pelagatti, P. 1976. Per il parco archeologico di Camarina, in B. d'A.: 122-141.

Pelagatti, P. 1977. Kaukana: un ancoraggio bizantino sulla costa meridionale, in Sicilia Archeologica, 18-1920, a.V., 89-99.

Pelagatti, P. 1978. Camarina, in Studi Etruschi, XLV: 571-574.

Pelagatti, P. 1979. Camarina, in Storia della Sicilia, I: 509-518.

Pelagatti, P. 1980. Attività della Soprintendenza alle Antichità per la Sicilia Orientale in Kokalas XXVI-XXVII II: 712-731.

Uggeri, G. 1961. Stazioni preistoriche costiere, in Note Camarinesi: 1-8.

Uggeri, G. 1974. Foglio 275 (Scoglitti), Edizione Archeologica della Carta d'Italia al 100.000.

ANNE-MARIE COLLOMBIER

Harbour or Harbours of Kition on Southeastern Coastal Cyprus

ABSTRACT

Kition was a major harbour town on the southeastern coast of Cyprus and lay in a particularly favoured area, within a large protected bay. The development and prosperity of the city depended on the copper industry and on sea-borne trade with the Aegean and the Near East.

Now the ancient site lies directly beneath the modern town of Larnaca, and is stranded inland as the coastline has advanced.

In this paper, we are dealing with the problem of the harbour town, from the Late Bronze Age to the end of the Phoenician period. All forms of evidence relating to the past physical and cultural environments of the settlement are considered.

Ancient excavations and more recent archaeological works in the northern (Kathari, area II, Cyprus Antiquities Department) and eastern (Bamboula hill, French Archaeological Expedition) quarters of the town established its importance as a port city of the Late Bronze Age, and produced new archaeological evidence for the Phoenician aspects of Kition.

The urban development of Kition may be followed more clearly. The Late Bronze Age town is situated in the northern area with its harbour, near Kathari. Then the town occupied the eastern side, where the well known harbour of Classical Kition is located. Thus, there are at least two different harbours in this archaeological site (without forgetting the neighbouring Late Bronze Age harbour at Hala Sultan Tekke, within a short distance to the south of Kition).

THE AREA OF KITION

Kition is situated on the western side of Larnaca Bay, extending from Cape Pyla to Cape Kiti, on the southeastern coast of Cyprus (Fig. 1). It lies in a particularly favoured area, within this large bay which

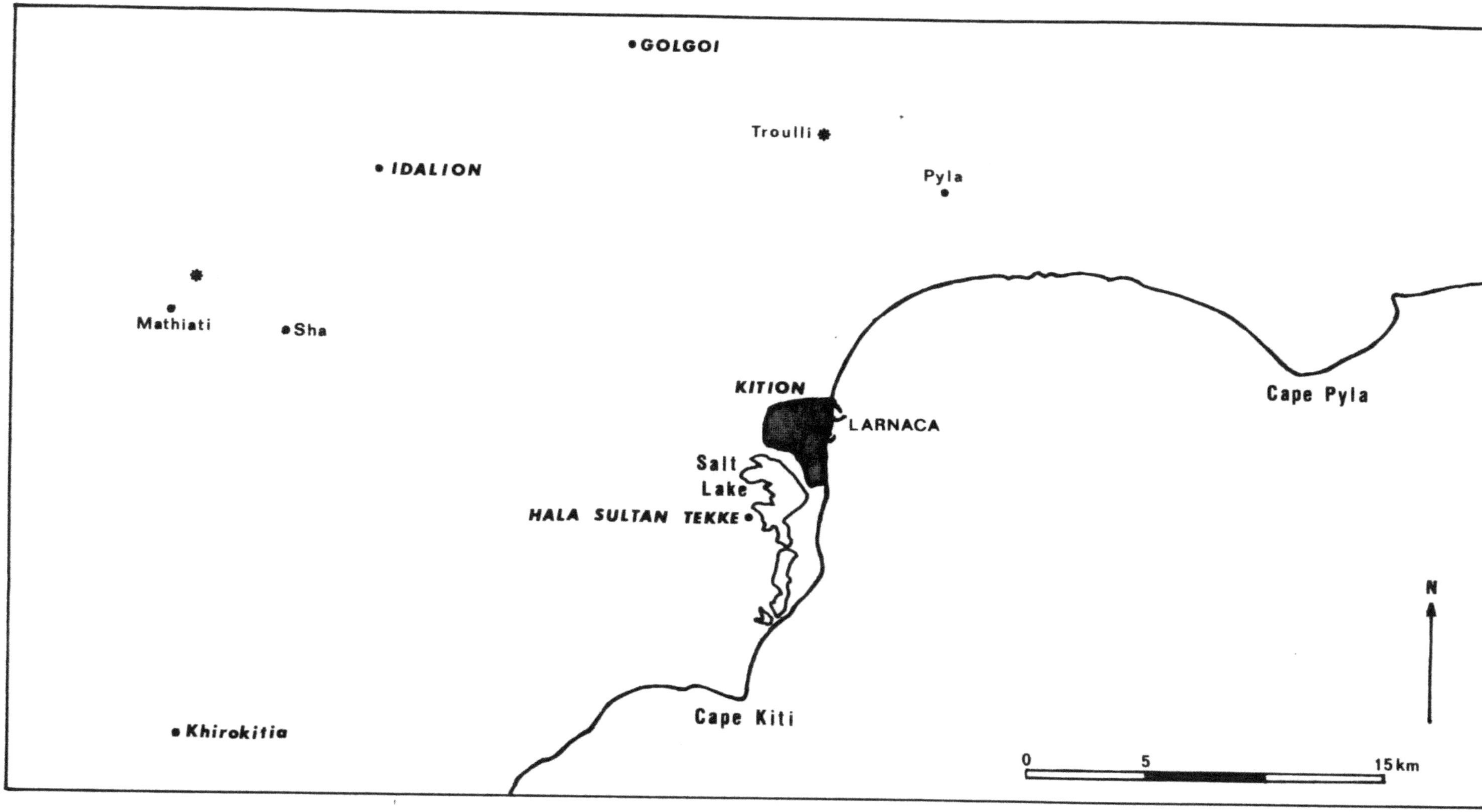

Figure 1. Area of Larnaca-Kition

provides shelter and safety. Protected from the strong winds and nearer to Egypt and the Syro-Palestinian coast than any other part of the island, this bay attracted the early Cypriots from the Neolithic period onwards (Catling 1966).

There is a fairly cultivable plain, the good fertile land being to the west and southwest of Larnaca, while the land north of the town is marshy. Here also is the neighbouring copper mine of Troulli, 12 km north of Larnaca, and northwestwards, the little farther copper ore of Sha-Mathiati (Bear 1963: 64-67; 76-95). Then, within a short distance to the south are the salt lakes of Larnaca. From the largest depression of the salina complex men would collect the salt just as it is still done today (Bear 1963: 175-179).

The environment of the area (fertile land, close proximity of the sea, good natural shelter, deposits of salt and copper ore) brought about the emergence of settlements along Larnaca bay, particularly near the west side of the Lake (Hala Sultan Tekke) and south of Pyla (Catling 1963: 149-168; Nicolaou 1976: 14, Fig. 3). Among them Kition became a major harbour town flourishing during Late Bronze Age and Iron Age. The development and the prosperity of the city depended on the copper industry and sea-borne trade with the Aegean and the Near East (Karageorghis 1976: 58-141; Karageorghis et al. 1981). Kition was also known to the biblical world as a maritime power (Hill 1949: 96-97; Gjerstad 1948: 436; 459; Bunnens 1979: 349-352).

At Kition we are faced with two main problems:

1. It has long been known that ancient Kition lies directly beneath the modern town of Larnaca, more precisely between Old Larnaca, a low plateau 500 metres inland, and Scala, on the sea-front (Nicolaou 1976: 9-51; 298-304). There are thus very few open spaces available for excavation. Moreover, since the events of summer 1974 the town has grown considerably and become densely inhabited, vacant areas quickly disappearing. Due to an almost complete cover of modern buildings and roads, the archaeological exploration and survey of Kition is difficult. It should also be stated that the site of Kition has been occupied at least since Early Bronze Age and has survived through centuries to the present day.

2. Larnaca region is a flat coastal plain, made up of sand, silt, and slightly consolidated shingle. The structure of this coastal plain is one of the beach deposits prograding eastward, sometimes trapping behind them depressions that become lagoonal environments (Gifford 1978: 56-84). The present coastline is well away from the ancient one (Fig. 2). The eastern limit of the archaeological site is now 600-500 metres from the shoreline. Kition once occupied a coastal setting and is now stranded inland. Here we have changes in shoreline configuration illustrated by archaeological evidence.

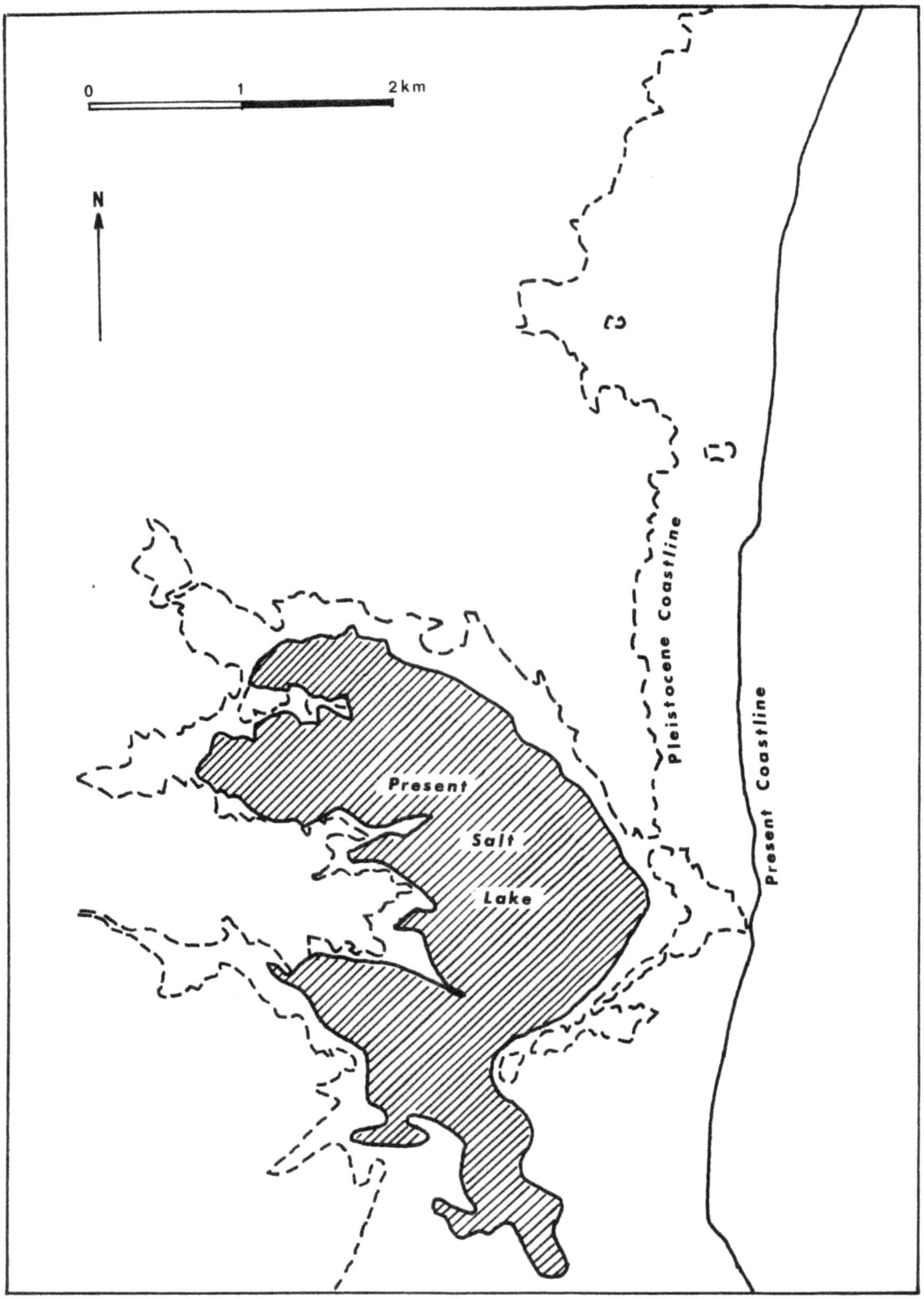

Figure 2. Evolution of the shoreline (after K. Nicolaou).

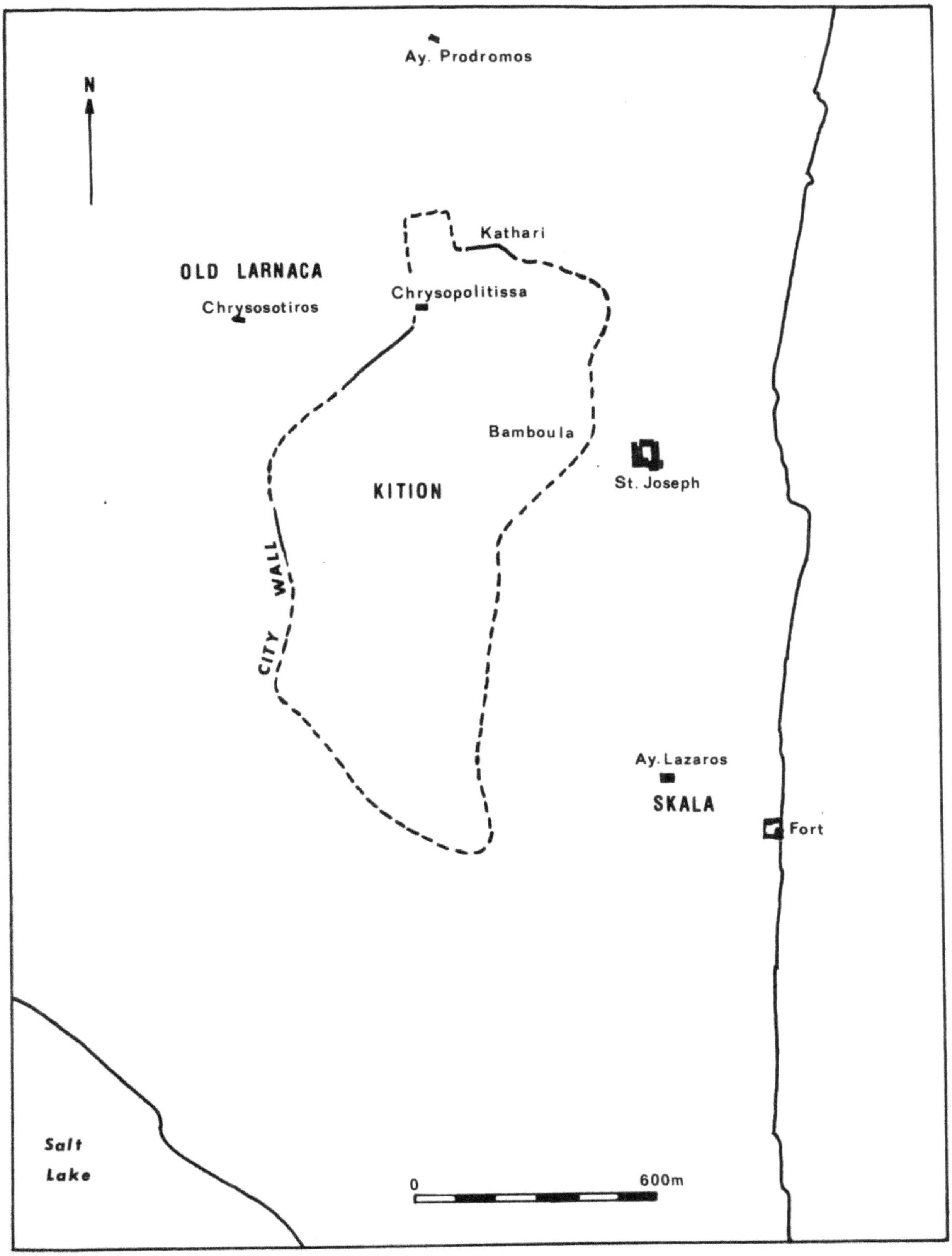

Figure 3. Map of Kition

THE ARCHAEOLOGICAL SITE

Kition has served as a profitable field for clandestine and amateur excavation over a long period in the past (Karageorghis 1976: 17-21). In 1929 and 1930, scientific excavations were carried out by E. Gjerstad but they were of a limited scope (Gjerstad 1937: 1-75).

New information has been brought to light by the latest discoveries of the Department of Antiquities of Cyprus. These comprise topographical and archaeological survey of the city published in 1976 by K. Nicolaou and systematic excavations at the localities of Chrysopolitissa and Kathari in the northern sector of the town, begun in 1959 and now completed by V. Karageorghis. (Karageorghis 1974; Karageorghis & Demas 1986).

Since 1976 preliminary reports and publications of the French Archaeological Expedition at Bamboula hill must be added (Calvet 1982; Salles 1983; Caubet & Yon 1985).

Kition can be traced within the modern town of Larnaca. The site is situated on a low plateau rising up to 10 metres above sea-level and which runs parallel to the present shoreline, about 500 metres away. This plateau is 1500 metres long and about 600 metres wide, the highest points are on the north by the Chrysopolitissa church and eastward at Bamboula hill (Fig. 3).

Archaeological evidence bears witness to the occupation of the site as early as the Early Bronze Age (Chamber tombs of the Early Bronze Age near the church of Ayios Prodromos, architectural remains and tombs near Chrysopolitissa church, c.1800 B.C.E.) down to the Middle Bronze Age. After that there is a period of about 400 years which is not yet represented archaeologically at Kition. No traces of the early phase of the Late Bronze Age have been recorded; this is probably due to chance.

During the Late Bronze Age - about 1300 B.C.E. - Kition was founded as a town. Since the archaeological material known so far comes from Kathari, Chrysopolitissa, and Bamboula hill it seems probable that the nucleus of the town occupied the northern part of the site. Archaeological evidence suggests different phases of the town and several waves of newcomers from the Aegean just before the end of the eighth century B.C.E. until the beginning of the eleventh century B.C.E. During the second quarter of the eleventh century B.C.E. a natural phenomenon, probably an earthquake, put an end to the Late Bronze Age town of Kition. After a short revival - c.1050 B.C.E. - several parts of the town were abandoned.

Between 1000 B.C.E. and the end of the ninth century B.C.E. there was a hiatus which has been interpreted as a result of the partial abandonment of the town. However by the ninth century B.C.E. there has been a revival on a large scale. Evidence of occupation over a wider area begins with the Geometric period and later. Kition spread beyond its original nucleus, mainly to the east and the south, with the Phoenician colonists from Tyre who founded and developed a new town (Guzzo Amadasi & Karageorghis 1977: 188-189; Masson & Sznycer 1972: 69-76).

THE HARBOUR OF HISTORIC KITION

Brief references in Thucydides (I, 114) and Strabo (XIV, 6, 3) indicate that Kition was a major harbour town. But Strabo only speaks of Kition's "kleistos limen", a harbour that can be closed, a harbour with an entrance "closable" at night. We do not know whether Strabo speaks of an enclosed harbour, i.e. within the city walls. However this information suggest that the harbour is not entirely or even partially natural.

This harbour has long since disappeared but its former position can be traced. From the sixteenth century to the beginning of the twentieth century the travellers who stopped at Larnaca on the pilgrim's road to the Holy Land, the antiquaries and the scholars who visited Larnaca recalled Strabo's mention of Kition's closed harbour and identified its remains in a marshy pool adjacent to Bamboula hill. Several maps published by these scholars clearly show the position of the harbour, as lying by the St. Joseph Convent and east to Bamboula hill (Nicolaou 1976: 27; 37-38; 41-43; 53; 56-57; 66).

A small channel 300 metres north to the St. Joseph Convent still runs from this area to the sea, but we do not know when it was built and eventually associated with the harbour (Cyprus Town Plans: Larnaca and Scala, grid: 647/579-580).

The extent of this harbour cannot be defined with certainty, but archaeological discoveries from ancient and new excavations have yielded more information as regards this quarter of the city. In 1879 in order to dry out the depression adjacent to Bamboula hill, British authorities filled up the marsh with the landfill taken from the hill. The field notes of Lt. Sinclair, in charge of the operations, give an indication of the nature of the buildings which were destroyed: ashlar blocks of stone, stone anchors and "two stones with holes through them were found fixed in the ground and apparently used for fastening the ships in the harbour" (Bailey 1969: 37; Ohnefalsch-Richter 1891: 478-481).

Several years before, L. Palma di Cesnola identified what he believed to be remains of the harbour installations around and beneath the St. Joseph Convent. He also recognized fragments of beach-rock in this area. (Cesnola 1877: 49).

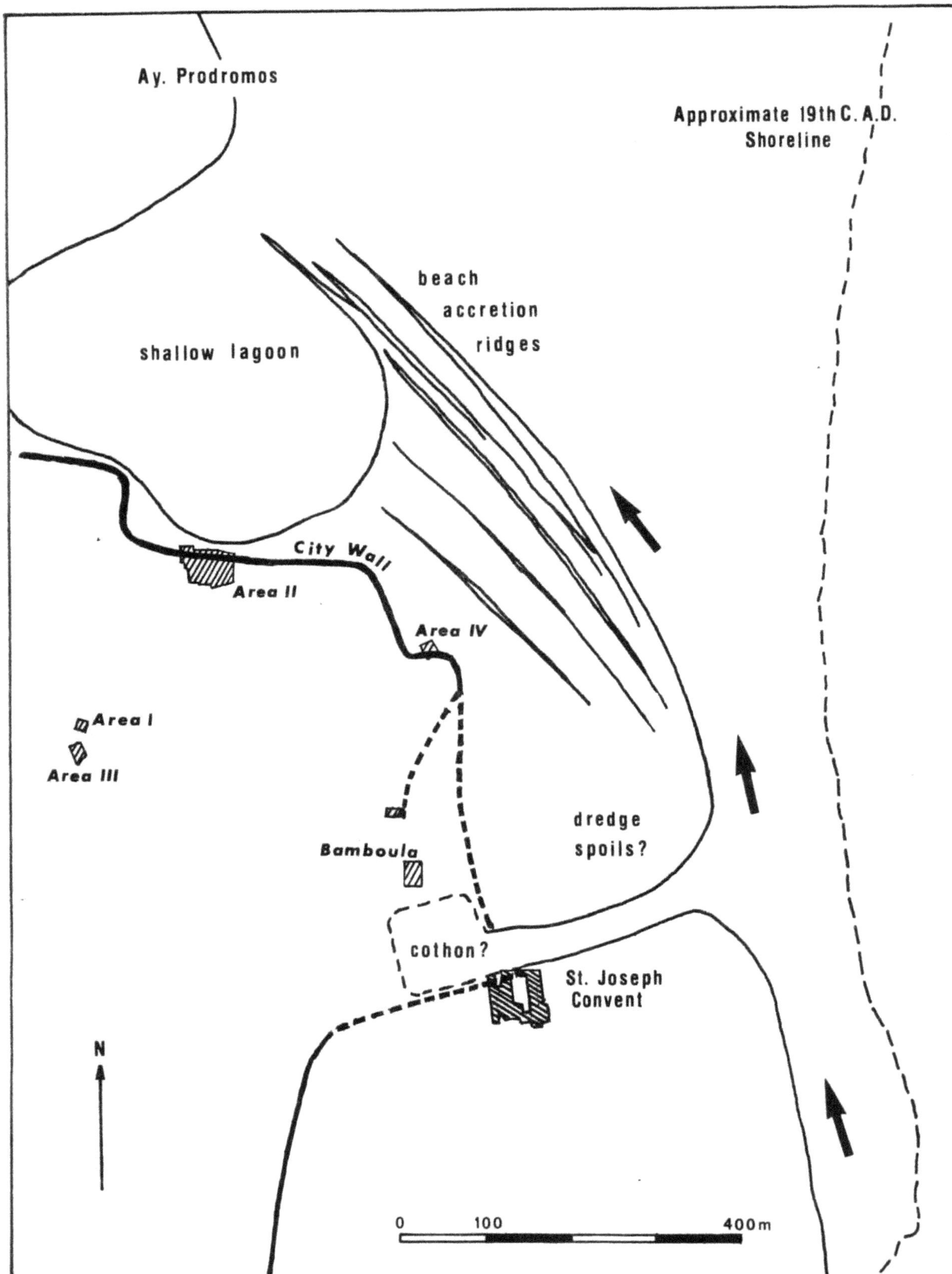

Figure 4. Palaeogeographic reconstruction of the coastal zone in the IVth century B.C.E. (from J.A. Gifford 1978: fig. 59).

The first results of the excavations at Bamboula produce new information:

(1) Bamboula hill is not the acropolis of Kition but an artificial mound raised in the Mediaeval period with the remains of a small fort (Yon 1984: 85). The mound, constructed with ballast taken from the adjoining area, contained many stamped amphora handles, from the fourth century B.C.E., to the second century B.C.E. which bear witness to storage areas and magazines near the harbour (Calvet 1982: 7-53).

(2) The area excavated supplies evidence of a sacred area with sanctuaries and altars from the tenth century B.C.E. to the Hellenistic period (Yon 1984: 85-97). The earliest layers are 2 metres above sea-level and Bamboula is a low, flat area near the sea-shore.

(3) One particular feature of the spot is the number of installations connected with the supply of water in the excavated area and northwards (Bailey 1969: 37; Nicolaou 1976: 145-155; Callot & Salles 1981: 54-57; 63-67). Artificially constructed water channels and drains trended eastwards. The water was running outside the town to the harbour.

A.J. Gifford established that about the fourth century B.C.E., the shoreline lay eastward of Bamboula hill (beach-rock associated with black-glazed sherds) and the palaeogeographic conditions would not produce a natural embayment in the vicinity of Bamboula (Gifford 1978: 67; 164). Gifford's observations and drillings north to St. Joseph Convent agree with Cesnola's remarks. Moreover in the northern side of the excavated area at Bamboula, the archaeologists found a section of the line of the city-wall (Caubet & Yon 1985: 10-11). If Gifford's hypothesis is correct we may suggest that the Phoenicians remedied the lack of natural harbour and constructed a "closed harbour" connected with the sea through a channel. (Fig. 4 with a different line for the eastern city-wall).

THE LATE BRONZE AGE HARBOUR

Topographical features and architectural remains may attest the existence of the harbour in the Late Bronze Age. In the northern part of the town (Kathari, area II) V. Karageorghis uncovered several sanctuaries of the Late Bronze Age, associated with metallurgical workshops. This area remained a sacred area until the end of the fourth century B.C.E. The Phoenicians selected the site of the largest of the earlier Bronze Age temples (Temple 1) and built a new temple dedicated to Astarte (?) on the foundations of the old one (Karageorghis 1976: 95-141).

The town never expanded beyond this northern limit. Several years ago, the nearby area north of the city wall was still marshy. The coreholes drilled by Gifford indicate a lagoonal environment at least until the Hellenistic period (Gifford 1978: 58-64). There was at least a natural embayment at the locality of Kathari.

Several dozens of stone anchors of various sizes, reused as building material (cyclopean wall, temples foundations) were found in the excavated area (Karageorghis 1976: 60).

Northwards beside the earlier city wall, the excavator brought to light two rectangular ashlar built structures and identified them with defensive bastions (Karageorghis 1976: 53-54; 60-61; pl.1, 30, 42 and XI). The orientation and the use of ashlars and paving slabs do not exactly correspond with the earlier brick wall of the town. Parallels and similarity with the harbour features of Dor established by A. Raban might make one think that these "bastions" were quays of the first harbour of Kition (Raban 1983: 238-241). This site was probably an inner harbour of Kition in the thirteenth century B.C.E. When people had to leave Hala Sultan Tekke's harbour sealed off from the sea, they moved to Kition headland. Kathari, where the sea penetrated, might have been used as a suitable place for a harbour.

REFERENCES

Amadasi, M.G. and V. Karageorghis. 1977. Fouilles de Kition III: Inscriptions Phéniciennes. Nicosie.

Bailey, D.M. 1969. The village priest's tomb at Aradhippo in Cyprus. Quarterly of the British Museum XXXIV, 1-2: 36-85.

Bear, L.M. 1963. The Mineral Resources and Mining Industry of Cyprus. Bulletin No. 1, Geological Survey Department. Nicosia.

Bunnens, G. 1979. L'Expansion Phénicienne en Méditerranée. Bruxelles-Rome.

Callot, O. and J.F. Salles. 1981. Un collecteur à Kition (Chypre). In: L'Homme et l'Eau en Méditerranée et au Proche-Orient. Travaux de la Maison de l'Orient No. 2. Lyon.

Calvet, Y. 1982. Kition-Bamboula I: Les Timbres Amphoriques. Paris.

Catling, H.W. 1963. Patterns of Settlement in Bronze Age Cyprus. Opuscula Ath. IV: 129-169.

Catling, H.W. 1966. Cyprus in the Neolithic, Chalcolithic and Bronze Ages. Cambridge Ancient History, Fasc. 43.

Caubet, A. and M. Yon. 1985. Kition-Bamboula III: Le Sondage L-N 13 (Bronze Récent et Géométrique I). Paris.

Cesnola, L. Palma di. 1877. Cyprus: Its Ancient Cities, Tombs, and Temples. London.

Gifford, J.A. 1978. Paleogeography of Archaeological Sites of the Larnaca Lowlands Southeastern Cyprus. University of Minnesota, Ph.D.

Gjerstad, E. 1937. The Swedish Cyprus Expedition, III. Stockholm.

Gjerstad, E. 1948. The Swedish Cyprus Expedition, IV: 2. Stockholm.

Hill, G.F. 1949. A History of Cyprus, Volume I: to the Conquest by Richard Lion Heart. Cambridge.

Karageorghis, V. 1974. Excavations at Kition I: The Tombs. Nicosia.

Karageorghis, V. 1976. Kition: Mycenaean and Phoenician Discoveries in Cyprus. London.

Karageorghis, V. 1959-1986: yearly. Chronique des fouilles et découvertes archéologiques à Chypre. Bulletin de Correspondance Hellénique, Paris.

Karageorghis, V., Coldstream, J.N., Bikai, P.M., Johnston, A.W., Robertson, M. and L. Jehasse. 1981. Excavations at Kition IV: The Non-Cypriote Pottery. Nicosia.

Karageorghis, V. and M. Demas. 1986. Excavations at Kition V: The Pre-Phoenician Levels. Nicosia.

Masson, O. and M. Sznycer. 1972. Recherches sur les Phéniciens à Chypre. Paris Genève.

Nicolaou, K. 1976. The Historical Topography of Kition. SIMA XLIII. Göteborg.

Ohnefalsch-Richter, M. 1893. Kypros, the Bible and Homer. London.

Raban, A. 1983. Maritime Archaeological Research in Israel. International Journal of Nautical Archaeology 12.3:229-251.

Salles, J.F. 1983. Kition-Bamboula II: Les Egouts de la Ville Classique. Paris.

Yon, M. 1984. Fouilles françaises à Kition-Bamboula (Chypre), 1976-1982. Comptes Rendus de l'Académie des Inscriptions, janvier/mars: 80-99.

FRANK J. FROST

A Submerged Settlement at Skyllaieis

Skyllaieis was a community of the ancient Greek polis of Troizen. Construction from the Classical period (5th-4th centuries B.C.E.) can be seen on the acropolis and a villa complex of the 4th century C.E. lies in the shallow water below the acropolis. About 600 m to the west of the villa a number of walls were found at a depth of 3m indicating a probable sea level rise since antiquity of 5m. The walls are primitive and asymmetrical and may be as early as the 5th century B.C.E. It is suggested that they are the remains of an isolated farmstead. 100m. farther west are three or more tile burials at a depth of about 2.5m.

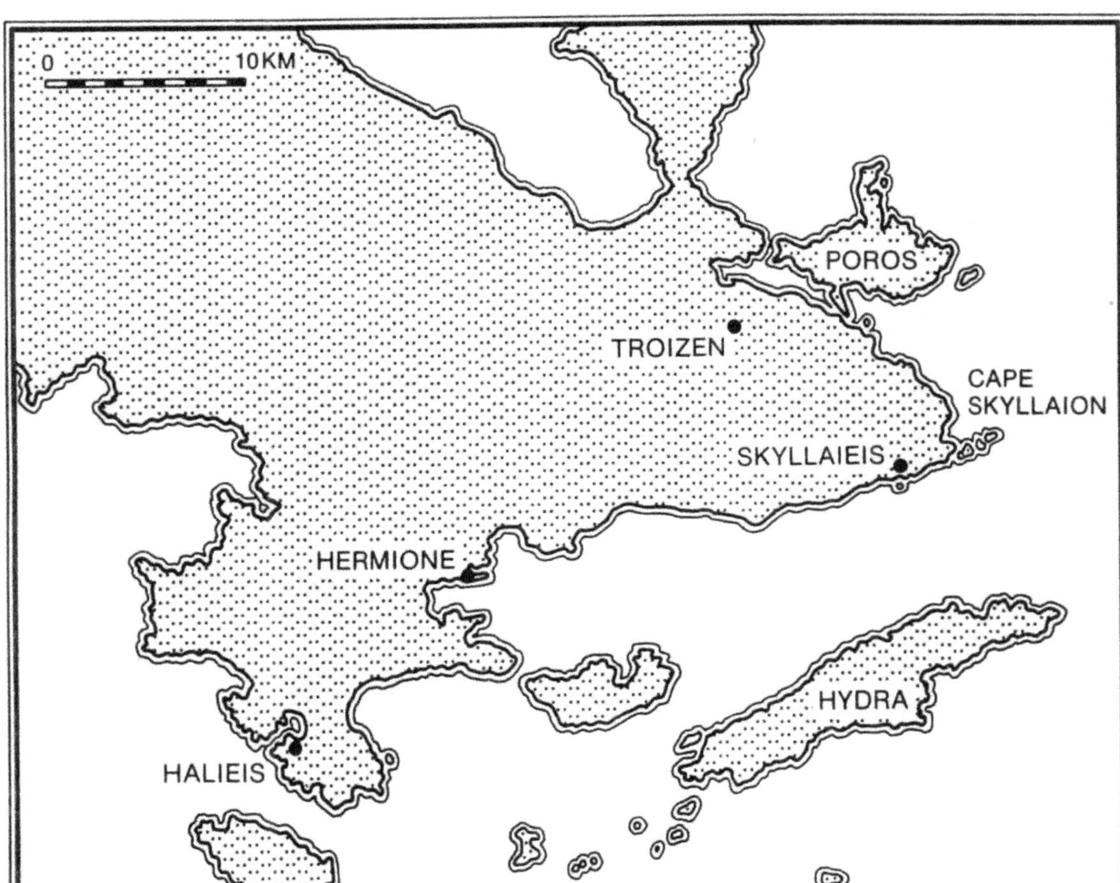

Cape Skyllaion was one of the major geographical reference points for maritime traffic in the ancient Greek world. All ships leaving Athens, Corinth, and other ports in the Saronic gulf for destinations in the

Argolid, the rest of the Peloponnesos, and points west would round the cape and generally take the more sheltered passage between the coast of the southern Argolid and the island of Hydra. Observers on such ships would have noted that the cape was followed by a number of steep cliffs descending into the sea, but that after a few kilometres the hills flattened out into a coastal plain that was obviously well watered and well suited for human occupation and cultivation. Fragments of obsidian and pottery sherds from all periods of the Bronze Age show, in fact, that the area was settled and no doubt farmed from the Neolithic. Just to the east of the plain is a picture book acropolis, which offered to the inhabitants of the plain a place of refuge from marauders. The small island of Soupia also provides some shelter for anchorages below the acropolis. All these advantages made the site attractive and it was probably sometime in the archaic period that the Dorian polis of Troizen, about 14 km to the northwest, organized it into a koma , or village community, and began to fortify the acropolis.

The Troizenians must have had ambiguous feelings about this outlying community. It offered good farming for Troizenian citizens, but its location was dangerously exposed. A fortified acroplis would serve to notify simple pirates sailing by that the area belonged to people who could defend themselves, but in a conflict between greater powers, it was an obvious target for plunder, as Thucydides makes clear during the early years of the Peloponnesian War, when the Athenians several times ravaged Troizenian territory (2.56.5, 4.45.2, and cf. 5.53).

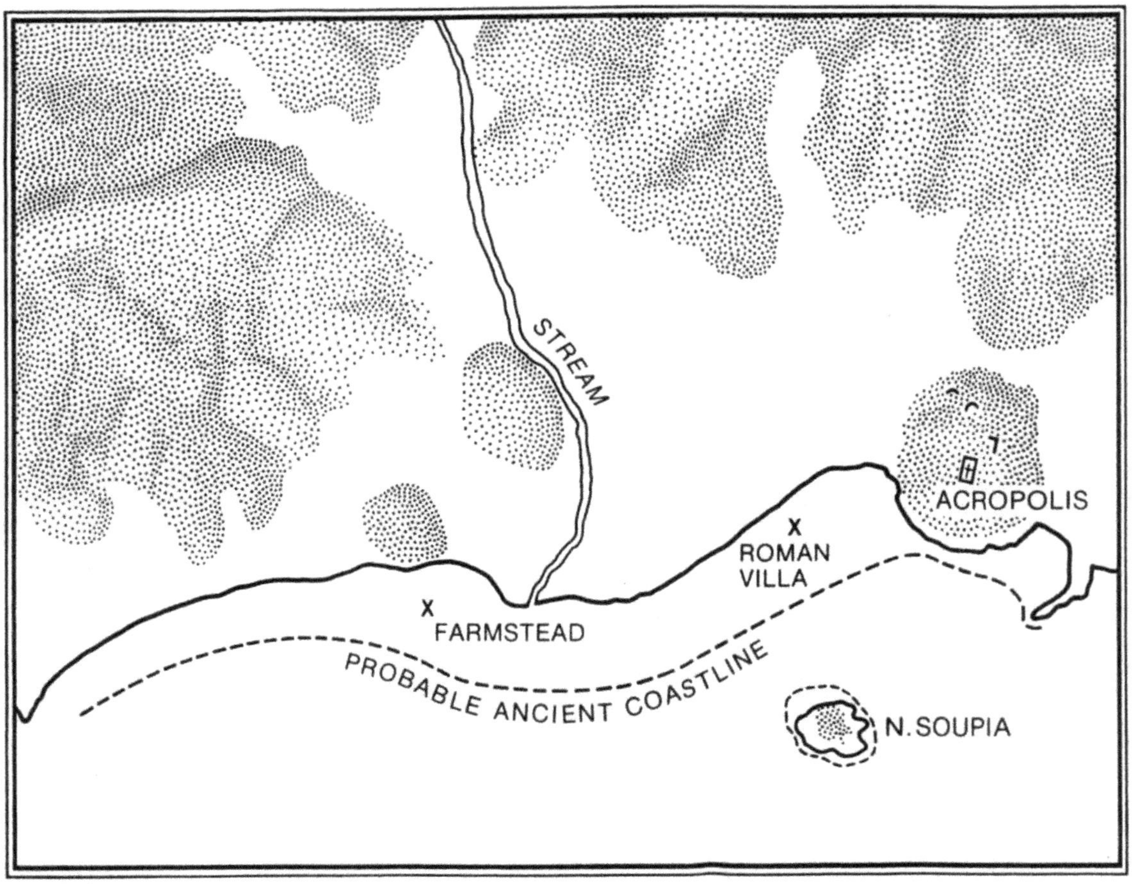

The history of Skyllaieis must be deduced entirely from archaeological evidence because our literary sources have never mentioned the site. Even the name Skyllaieis is a conjecture. The modern name is Phourkari, derived eventually from the Latin word furca or "fork". At one time I believed the name to go back to one or another of the Latin meanings. But it was recently drawn to my attention that there is a Vlach village named Fourka in northern Greece and that the word fourka, in the Romanian dialect used by the Vlachs, means "distaff" (Hammond, 1976). Because there are many Vlach settlements in this part of the Argolid, the name Phourkari must be Vlach and relatively recent. Now when the ancient name for a place has been forgotten and our ancient topographical writers do not provide a name, identification depends on the chance find of an inscription that names the place. Excavation may some day unearth such an inscription; fortunately we have a decree from Troizen in the Hellenistic period that names various groups of Troizenians who are to be taxed to pay for new fortifications. One of these groups is named as Skyllaieis, a name which would seem to refer to Cape Skyllaion. Thorough reconnaissance of the entire area demonstrates that only the fields and acropolis of Phourkari could have provided the site of an ancient village in the vicinity of Cape Skyllaion (Frost, 1980, Maier, 1959).

Masses of pottery sherds testify to long and continuous occupation. Other than a very few sections of foundation here and there, however, the only construction from the Classical period is found on the acropolis. The visitor will notice first that some of the walls and foundations of the mediaeval monastery of Ag. Athanassios were built partially with masonry from Classical walls. On the northern slope of the acroplis are two round towers and the northeastern corner of the city wall remains well preserved.

The most obvious ruins other than those on the acropolis are walls leading out into the shallow water of the bay. In 1974, with the permission of the Greek Archaeological Service, we mapped what seemed to be the most substantial buildings (Frost, 1977). Tile and mortar construction indicated a date sometime in the later Roman empire and the final plan seemed to show a villa in a courtyard, surrounded by other buildings of unknown purpose. Slightly to the west was a small bathhouse with an apsidal foundation. Although most of the associated pottery has long since been lifted by souvenir-hunting snorkelers, we did find enough of one amphora to compare with an unpublished example from the Athenian agora, which can be dated to the later 4th century of our era. The villa complex then was built in a period of relative security, between the Heruli raids of the 260s and the invasion of Alaric at the end of the 4th century.

The territory of Skyllaieis is cut by a stream that flows only in winter and spring. Because neither pottery nor ancient construction seemed to continue west of the stream bed, the land survey made only a cursory search of the western fields and beach. It was therefore only a lucky glimpse from a small airplane flying from Porto Cheli to Athens that led us to a collection of walls west of what must have been the main area of settlement. In the fields west of the stream rise two hills. The westernmost one rises gradually to a gently rounded summit but ends abruptly where cliffs plunge into the sea to interrupt the line of the beach. Unlike the walls in the area of the villa, there are no traces of construction on shore to provide a clue to more buildings in the sea.

There were other indications that the walls had no connection with the villa and bath of the Roman period. They contained no mortar and were far more primitive. They were also considerably deeper, which made progress in mapping slow, as we had come equipped to measure walls in waist deep water, not in water to a depth of 3m that we found here in the western sector. The tops of what construction remains are from 2.5 to 3m. submerged. If one assumes another metre for the height of the walls and approximately one more to keep the footings above the ancient water table, one can demonstrate at least a 5m rise in sea-level since the day the walls were built. Now the villa of the Roman period, which was less than 1m submerged, was probably at or near the water's edge when it was built. This seems to have been the usual practice, as shown by a similar bath built on top of the classical walls at Halieis 30 kilometres to the west (Jameson, 1969). The greater depth of these walls therefore may confirm that they are older, although the case is still entirely circumstantial.

The plan also reveals a primitive design. With the exception of the smaller building to the east, the walls are built at haphazard angles of stones crudely piled on top of each other. There are no visible roof tiles and only a few pieces of coarse pottery that seem to be undatable. One is reminded of current usage in the modern Greek countryside, where walls tend to be as much a byproduct of clearing stones from fields as anything else. They are then used to delineate property lines, to build crude shelters, and to pen animals. Because the compound lies beyond any possible fortifications it may predate the

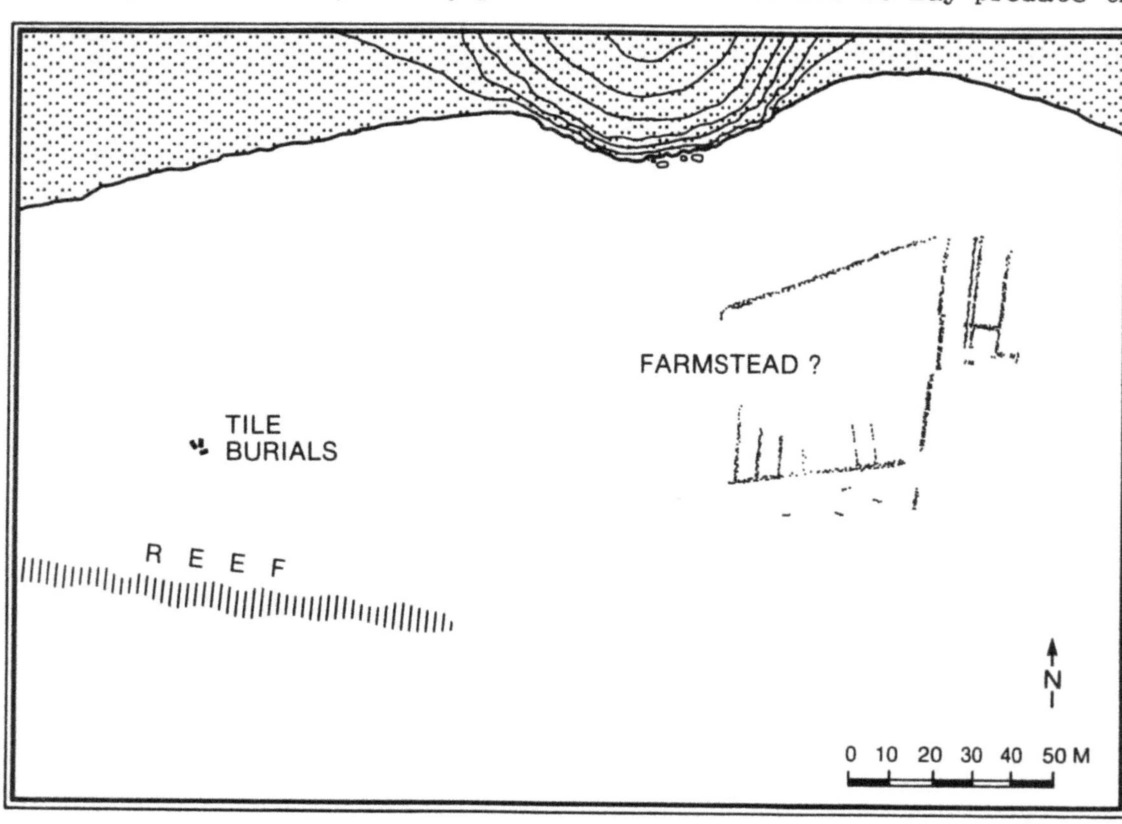

heightened naval hostilities of the 5th century, for it would certainly have offered an exposed target for any maritime marauders sailing by. My working hypothesis is that the structures are part of an isolated farmstead on the very outskirts of Skyllaieis, one that was possibly deserted because it was too exposed, or perhaps even because of a rising sea-level.

About a hundred metres even farther to the west is a small group of burials about two and a half metres below the surface of the sea. The graves are quite simple, lined with small stones and covered with roof tiles, a method of interment that does not predate the late archaic. It is not known whether the graves contained pottery or other goods and it probably never will be. On a recent visit I noticed that local bathers had been snorkeling around the graves and had pulled up some of the tiles out of curiosity. One hopes that there are some graves deep enough under the silt to escape notice.

About 20m farther out to sea from the burial site, the water begins to shallow as a low ridge rises from the sea bottom, roughly parallel with the shore. On the seaward side the ridge falls away sharply to greater depths. The area of the cemetery did not overlook the shore, therefore, but lay in a shallow valley between the hill and ridge to the north, and the lower ridge to the south. As the sea-level rose, wave action removed an unknown depth of soil from the ridge and invaded the valley. The appearance of the terrain was thus quite different in classical antiquity.

Too often in the past, excavation in the Greek world has been influenced by what we know from literary testimony. Excavators tend to go immediately for the monumental architecture, inspired by the texts of the classical historians. It seems to me that Skyllaieis offers a different opportunity - a chance to excavate an ancient Greek koma with absolutely no preconceptions about its history and to reconstruct the way of life of the complete spectrum of its citizens, from leaders and priests to the simplest farmers in its fields.

REFERENCES

Frost, F.J. 1977. Phourkari. A Villa Complex in the Argolid. International Journal of Nautical Archaeology 6: 233-238.

Frost, F.J. 1980. Skyllaieis, a District of Troizenia. American Journal of Archaeology 84: 186-188.

Hammond, N.G.L. 1976. Migrations and Invasions in Greece and Adjacent Areas. Park Ridge, New Jersey, p. 40.

Jameson, M. 1969. Excavations at Porto Cheli . Hesperia 38:239.

Maier, F.G. 1959. Griechische Mauerbauinschriften I. Heidelberg, no. 32, pp. 139-145 (= Inscriptiones Graecae IV. 757ab. 24-25).

GERD HOFFMAN, HORST D. SCHULZ

Coastline Shifts and Holocene Stratigraphy on the Mediterranean Coast of Andalucia (Southeastern Spain)

ABSTRACT

In a joint program with the German Archaeological Institute Madrid, the Fachbereich Geowissenschaften of the University of Bremen investigated the stratigraphy of Holocene sediments in the valleys of the Mediterranean coast of Andalucia. The study area of investigation is situated between the Rio Almanzora (province of Almeria) in the east and the Rio Guadarranque (province of Cadiz) in the west (Fig. 1). The aim of this cooperation between different sciences is the investigation of human settlements and the reconstruction of the ecological environment of this area, especially in historical times. A particular interest of this project is the evolution of coastlines during the Holocene period.

The investigations began in the valley of the Rio de Velez near Torre del Mar, approximately 30 km east of the town of Malaga. The first results showed a marine transgression into the valley in the early Holocene and a constant sedimentation with a rate of 1 to 2 m per millenium until the end of the 15th century C.E. After this period, when the Catholic kings of Spain expelled the Moors from the southern part of the Iberian peninsula, an enormous increase in the sedimentation rate was caused by the deforestation of the hinterland. The sediments of the 16th and 17th centuries have a thickness of more than 15 to 20 meters. During this time the valley was filled up with the eroded soils of the hinterland and the river Rio de Velez developed a delta.

The dominant geomorphological factor for this evolution of landscape was man, supported by a wetter and cooler climate at the end of the Middle Ages and the unstable equilibrium of the forests in the Mediterranean subtropics.

These results initiated the excavation of archaeological beds of the Phoenicians near the old coastline north of the Phoenician manufactory Toscanos. So the geological stratigraphy of the Holocene sediments could be correlated with the archaeological stratigraphy of the excavation.

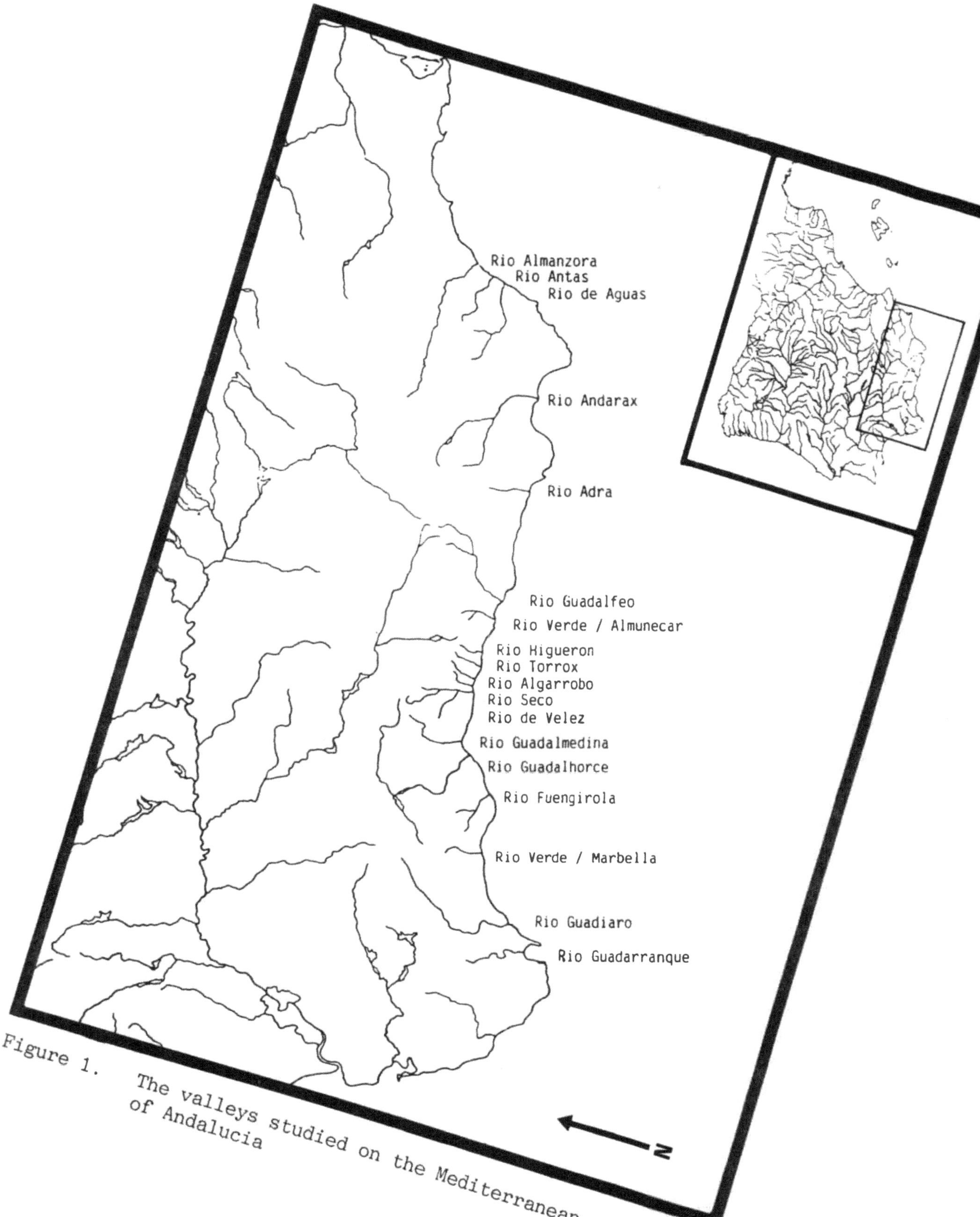

Figure 1. The valleys studied on the Mediterranean coast of Andalucia

INTRODUCTION

In an interdisciplinary project with the German Archaeological Institute Madrid, the Fachbereich Geowissenschaften of the University of Bremen investigated the stratigraphy of Holocene sediments in the valleys of the Mediterranean coast of Andalucia with various groups of students. The study area is situated between the Rio Almanzora (province of Almeria) in the east and the Rio Guadarranque (province of Cadiz) close to Gibraltar (Fig. 1). The aim of this joint program between geology and archaeology is the investigation of the relationship between human settlements and the evolution of the landscape. In regard to the reconstruction of the ecological environment, the changes in the coastlines during the Holocene is of particular interest to the project (Arteaga, Hoffmann, Schubart, Schulz 1986).

THE VALLEY OF THE RIO DE VELEZ (PROVINCE OF MALAGA)

The investigations started in the valley of the Rio de Velez near Torre del mar (Fig. 1), approximately 30 km east of the town of Malaga (Schulz 1983; Schulz, Jordt, Weber 1986). The reason for starting with this valley was that the German Archaeological Institute had excavated numerous Phoenician settlements in this region, in particular close to the mouth of the Rio de Velez (Fig. 2). The Phoenicians settled in this area between the 8th and the 6th century B.C.E. (Schubart 1982).

The most important settlement in regard to the geological investigations is the Phoenician manufactory Toscanos (Fig. 2), situated about 1 km from the recent coast on a little hill near the western valley slope. In the small bay north of this settlement the German Archaeological Institute Madrid excavated the Phoenician archaeological beds just beside the old coastline in 1985.

The geological situation in regard to the investigation of the Holocene stratigraphy is as follows: A large Holocene valley plain is framed by metamorphic rocks of the Betic Cordillere and sedimentary rocks of Tertiary and Pleistocene age. The valley floor has a length of nearly 10 km and an average width of 1 km (Fig. 3).

HOLOCENE STRATIGRAPHY

For the investigation of the Holocene sediments we used a hand auger of the Eijkelkamp Company (Netherlands). With this hand auger a maximal depth of 19.5 m was reached. Auger bits with a diameter of 10 cm were mostly used.

The grain size of the typical Holocene sediments in this valley is a sandy silt or a clayish silt. Only a small quantity represents middle and coarse sands or gravel.

Figure 2. Phoenician and Roman settlements close to the mouth of the Rio de Velez (Arteaga 1985). The arrow marks the "harbour bay" of Toscanos (Fig. 9).

The most striking macroscopic feature of the sediments is that at a certain depth the colour changes from brown to bluish grey. The colour shows the oxidation state of the iron and consequently the oxidation state of the sedimental environment: the brown colour testifies an oxidated, terrestrial environment and the bluish grey to an aquatic environment. In Figs. 4 and 5 some of the drilling profiles are drawn in relation to recent sea-level. It is striking that the change in colour from brown to bluish grey is quite constant, occurring at a depth close to recent sea-level. Most of the sediments of reduced bluish grey colour have been investigated micropalaeontologically. Many marine and brackish microfossils were found. These sediments are proof of an old marine or brackish bay, which had been filled up and covered by brownish flood plain deposits of the Rio de Velez.

DATING METHODS

Two methods were used for dating the Holocene sediments: (a) the radio-carbon method and (b) dating with ceramic fragments that were found in the drilled sediments.

The importance of dating the Holocene sediments by artefacts must be emphasized, especially in connection with the reconstruction of the ancient landscape. Not only is dating with this method usually more exact than a radiocarbon date, but the lack of organic material in the sediments of the Mediterranean valleys in this area underlines its importance. For saving the ceramic fragments it was necessary to use the auger bit with the diameter of 10 cm.

EVOLUTION OF LANDSCAPE

The east-west lithological profile (Fig. 6) near the settlement of Toscanos is constructed using the results of seven drillings, marked as vertical lines. In the lower part of the profile a clayey silt was found, covered by a sandy silt. Both sediments are bluish grey. The sediments above sea-level, the flood deposits, are also sandy silts. It is not possible to distinguish them lithologically from the layer below. The typical river sediments like sand and gravel are intercalated in the flood deposits. Near the slopes of the valley archaeological beds and allochthonous ceramic fragments were found.

By radiocarbon dating and by ceramic fragments, dated by Arteaga, we can date the top of the clayey silt, which reaches to 3 m below the recent sea-level, 1400 C.E. Prior to this period a marine or brackish bay existed in this valley (Figs. 11 and 12). Radiocarbon dates of organic material found in this clayey silt prove a sedimentation rate of these marine sediments of between 1 and 2 m per millenium, which is quite normal for marine sediments of this type.

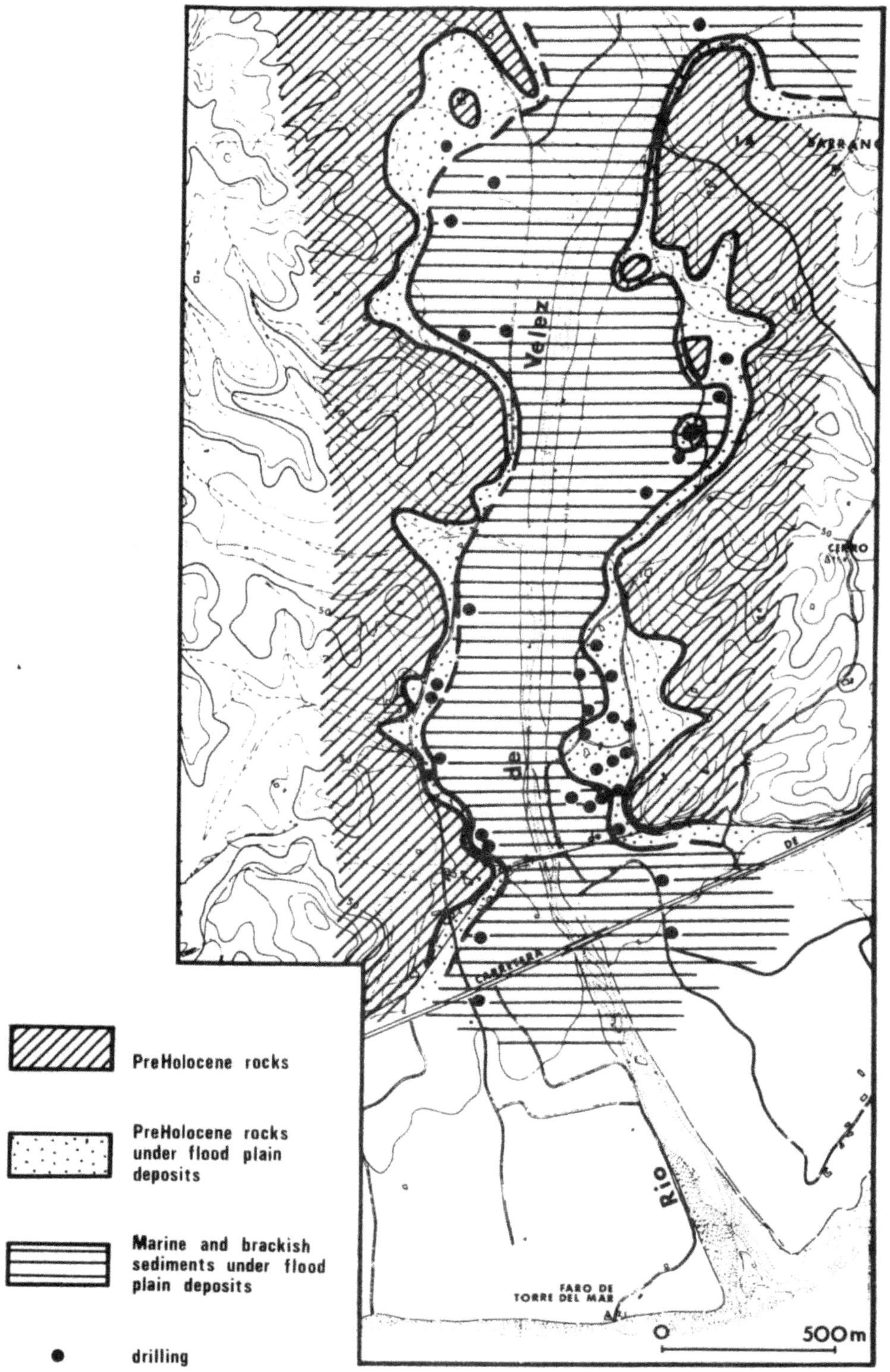

Figure 3. Occurrence of Holocene sediments in the valley of the Rio de Velez.

Around the 14th century we note an enormous increase in the sedimentation rate, up to nearly 12 to 15 m per millenium. This increase must have been initiated by a deforestation in this area during the Reconquista period, when the Catholic Kings of Spain expelled the Moors from the southern part of the Iberian peninsula. These very young deposits also containing coarser components like sand and gravel, filled up the valley in a very short time (Fig. 13). It is known that the ships of the Catholic kings sailed into this bay about 1450 C.E. to conquer the town of Velez-Malaga, which is now situated about 5 km from the coast (Figs. 3 and 12). Until the Reconquista, Velez-Malaga was a fortress of the Moors. These very rapidly deposited sediments are also noted in the schematic north-south profile (Fig. 7).

There are two possible explanations for the enormous increase of the sedimentation rate, either climate or man as the dominating geomorphological agent. Man's impact on the evolution of the environment in this region is documented by the deforestation during the Reconquista period, which was definitely followed by an enormous erosion of the soils in the catchment area of the Rio de Velez.

The natural climatic origin for this ecological catastrophe could have been the "little glaciation" (1550 - 1770 C.E.) with an increase in the rainfall of this region, which would also support the change in the erosion rate. But we know, that in earlier times during the Holocene there were several periods with very similar climatic conditions, but without such a large-scale erosion.

So man was the dominant geomorphological factor. The "little glaciation" only accentuated the erosion by an increase in the rainfall.

THE EXCAVATION OF THE "HARBOUR BAY" OF THE PHOENICIAN SETTLEMENT OF TOSCANOS

After the geological investigation in the valley of the Rio de Velez the German Archaeological Institute Madrid carried out an excavation in the "harbour bay" north of the Phoenician settlement of Toscanos (Fig. 2).

The intention of this excavation was to look for archaeological beds of the Phoenicians near the old coastline (Schulz 1986). In order to locate the excavation site exactly the old coastline in this little bay had to be accurately mapped. For this reason 62 handdrillings were first carried out (Fig. 8). With the results of these drillings it was possible to reconstruct the former landscape of this bay very exactly.

The surface of the ancient landscape in this little bay, is drawn in Fig. 9 using the method of krigeage (Matheron 1963). With the geological and archaeological results of the excavation of the "harbour bay"

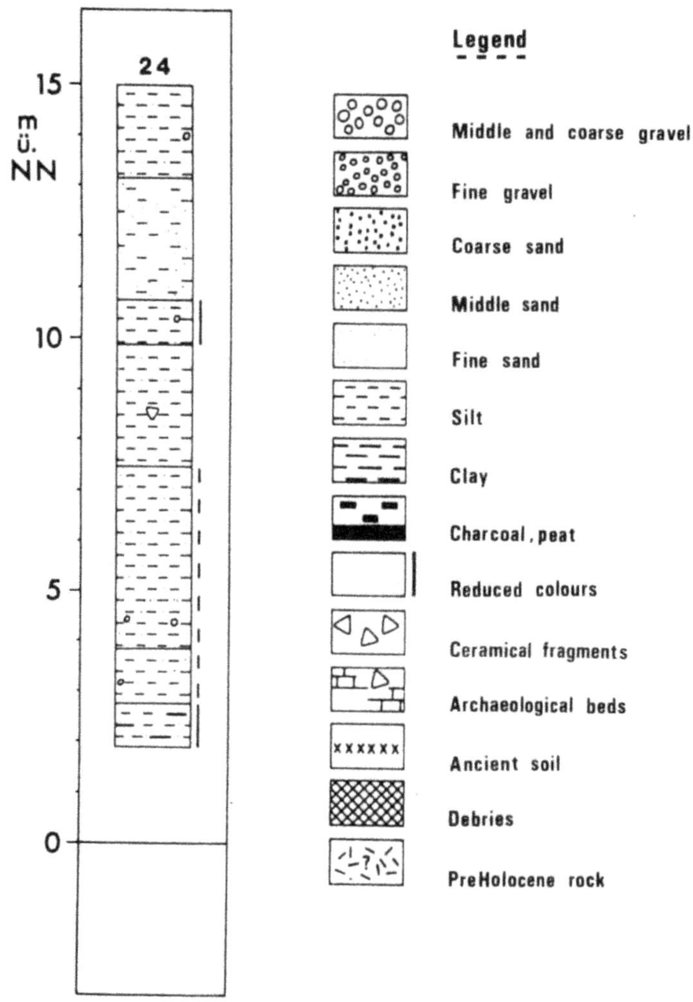

Figure 4. Drillings in the valley of the Rio de Velez.

of Toscanos it was possible to correlate the geological stratigraphy of the Holocene sediments with the archaeological stratigraphy of the excavation in an isochronous profile (Fig. 10). This profile is oriented in an east-west direction and reaches from the excavation to the centre of the valley. The lines are surfaces of the same age. The age of the ancient surface (Figs. 9 and 10), built by the slates of the Betic Cordillere, was presumed to be approximately 10,000 years. At this time the marine or brackish sedimentation in the bay began (Fig. 11).

In the 8th century B.C., when the Phoenicians settled in Toscanos, the bay was partly filled up, so that in the centre of the bay the water was still 6 - 7 m deep. The metamorphic slates at the shore of the bay had not yet been covered by terrestrial sediments. This is proved at least for the little bay of Toscanos, because the archaeological beds of the Phoenicians were found directly on the slates. These autochthonous beds of a settlement found in the excavation correspond with a very small layer of marine sediments in the centre of the bay, deposited during the same time.

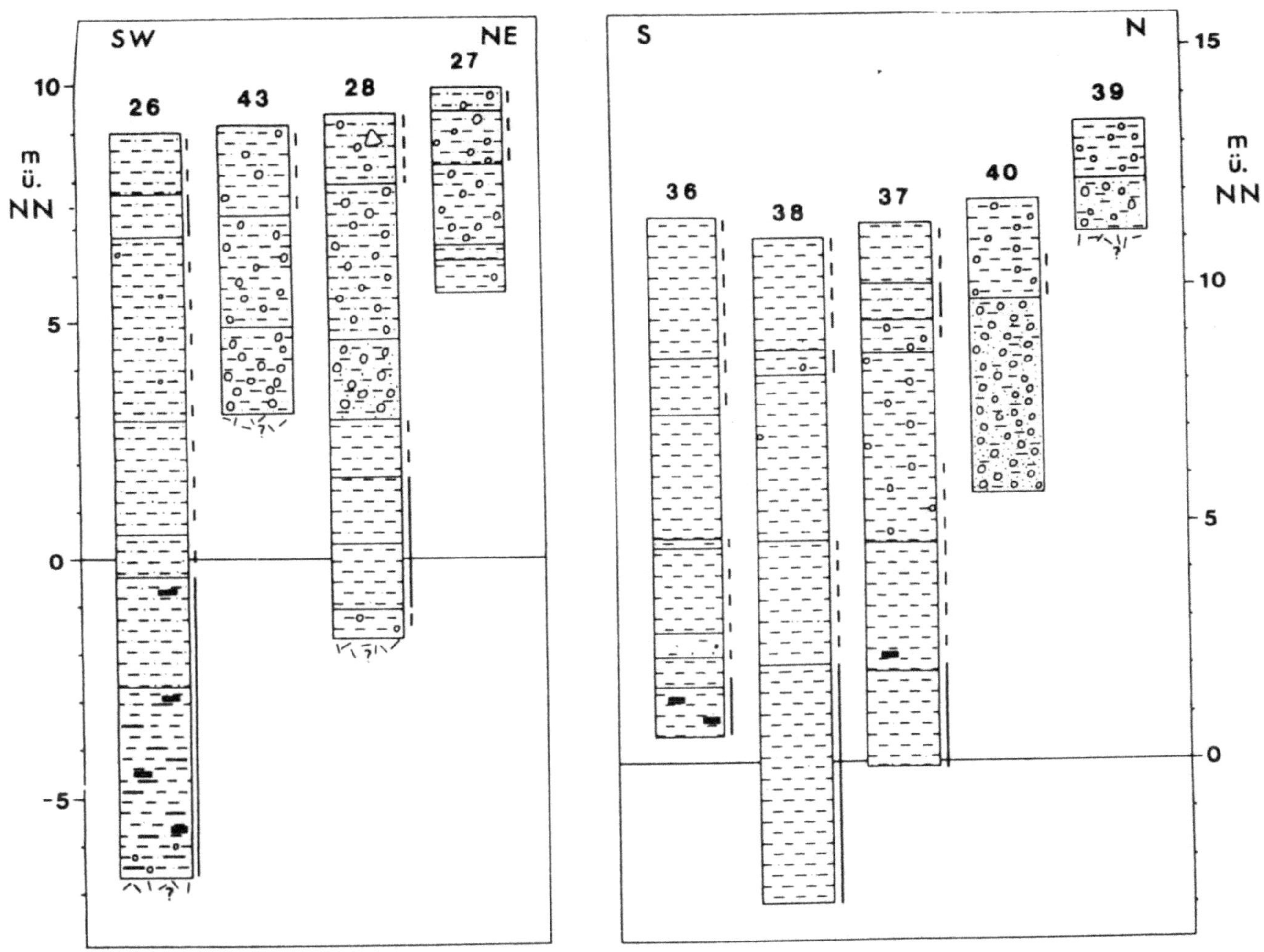

Figure 5. Drillings in the valley of the Rio de Velez.

The next well dated layer of the excavation is the surface of a Roman necropolis which is marked in the profile with the isochrone of 400 C.E. This line crosses the sea-level at a distance of about 40 m from the rocky shore. This fact and the composition of the sediments indicate that in late Roman times a sandy silt belt already existed on the shore of the bay.

The next well dated horizon of the excavation is the base of the very rapidly deposited sediments after mediaeval times, dated at 1400 C.E. So the bed laid down between 400 C.E. and 1400 C.E. is very thin. During this period nobody lived in the area of Toscanos. The sediments are flood deposits of the river and eroded soils of the abandoned settlement of Toscanos.

The prolongation of the 1400 C.E. isochrone corresponds with the centre of the bay with a water depth of about 3 metres. Still it was possible that the ships of the Catholic Kings could sail into the bay to attack Velez-Malaga. Near Toscanos the silty coastline was situated at a distance of about 100 m from the rocky shore of the Phoenician period.

The coarser sediments at the top of this isochrone filled up the bay within 300 years between 1400 and 1700 C.E. The marine and brackish sediments were covered by flood plain deposits and the river delivered a delta.

Between the 1700 C.E. isochrone and today only flood plain deposits have been laid down. A coarser layer of very sandy silt is correlated with a well known catastrophic flood about 100 years ago. For this period we can establish a sedimentation rate of 7 - 8 m/1000 years. This rate is lower than the rate during Mediaeval times, but higher than that before Mediaeval time. One reason for this fact is the lack of vegetation in the hinterland of the Rio de Velez in this period until today, an important difference between Phoenician and modern times.

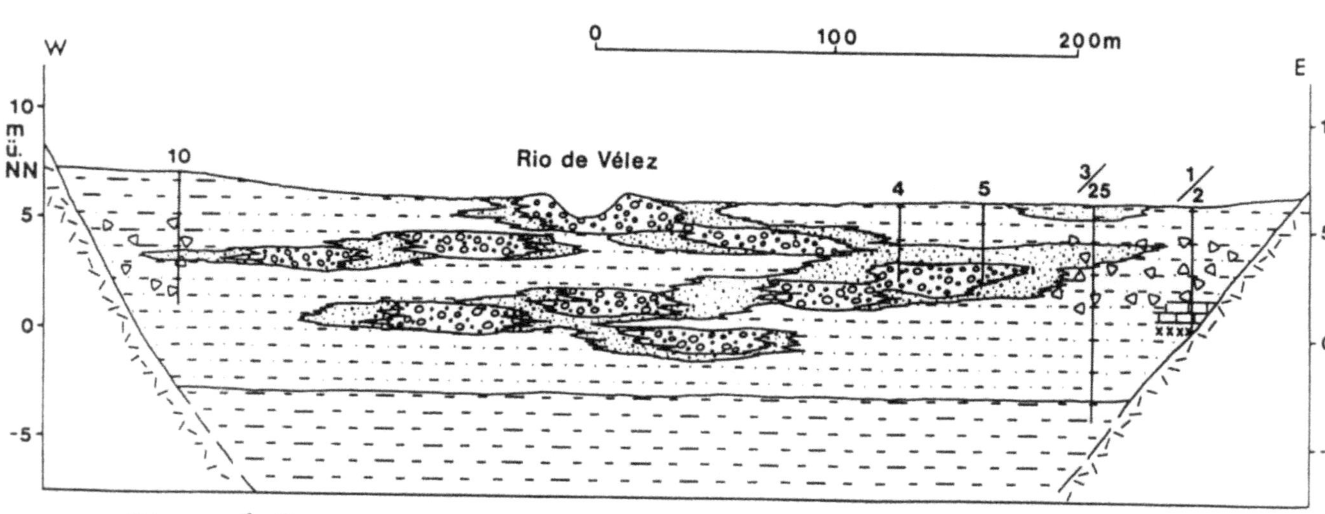

Figure 6. East-west lithological profile near the settlement of Toscanos.

SUMMARY

The correlation of the geological and the archaeological stratigraphy in the valley of the Rio de Velez proves a detailed reconstruction of the changing coastline in this valley. The evolution of landscape must be seen in the context of the climatic vegetational, morphological and especially human influences. The forests of the Mediterranean subtropics are located between the evergreen forests and the steppe. Because of this border situation, these forests have always been in disequilibrium. A climate with winter rains following a dry summer period, sometimes with torrential rainfall, supports the morphodynamic influence (Tichy 1962). Due to this ecological situation, man's impact on the ecosystem has been catastrophic in the Mediterranean subtropics, as the example of the Mediterranean coast of Andalucia shows.

The project is supported by the Volkswagen foundation.

REFERENCES

Arteaga, O. 1985. Los hornos romanos del Manganeto, Almayate Bajo (Malaga). Noticiario Arqueologico Hispanico, 23: 177-193.

Arteaga, O., Hoffmann, G., Schubart, H., Schulz, H.D. 1986. Geologisch-Archäologische Forschungen an der andalusischen Mittelmeerküste. Madrider Mitteilungen 27.

Matheron, G. 1963. Principles of Geostatistics. Economic Geology 58: 1246-1266.

Schubart, H. 1982. Phönizische Niederlassungen an der Iberischen Südküste. Madrider Beiträge 8: 207-234.

Schultz, H.D. 1983. Zur Lage holozäner Küsten in den Mündungsgebieten des Rio de Velez und des Rio Algarrobo (Malaga). Madrider Mitteilunten 24: 59-64.

Schulz, H.D. 1986. Geologische Bearbeitung der Hafenbucht Toscanos. Madrider Mitteilungen 27.

Schulz, H.D., Jordt, K.P., Weber, W. 1986. Stratigraphie und Küstenlinien im Holozän (Rio de Velez, Andalusien). Madrider Mitteilungen 27.

Tichy, F. 1962. Die Wälder der Basilica und die Entwaldung im 19. Jahrhundert. Heidelberger Geographische Arbeiten 8.

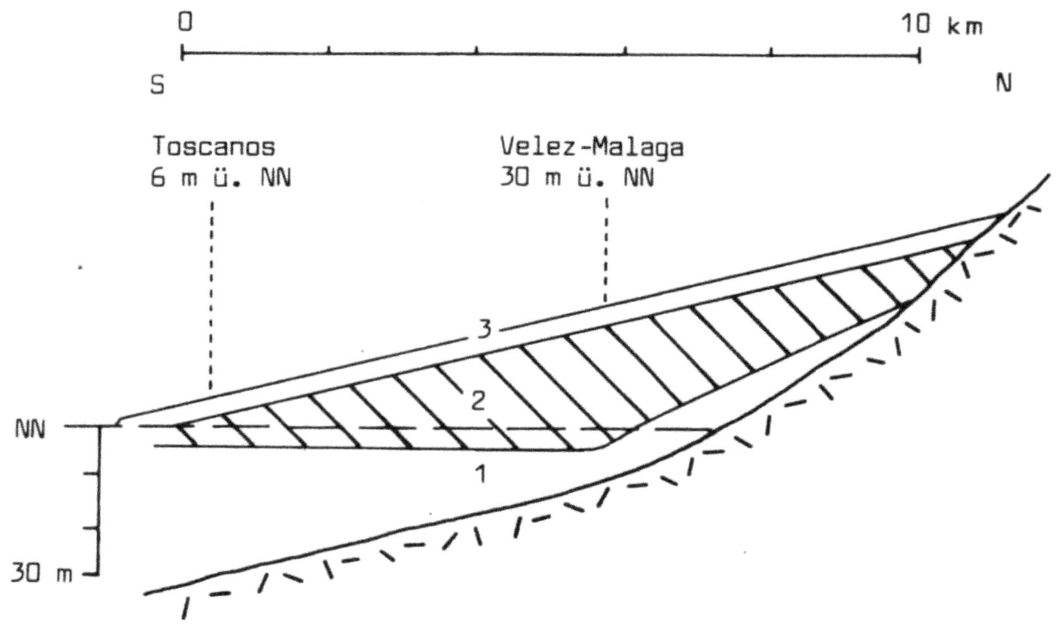

Figure 7. Schematic north-south profile of Holocene sediments.

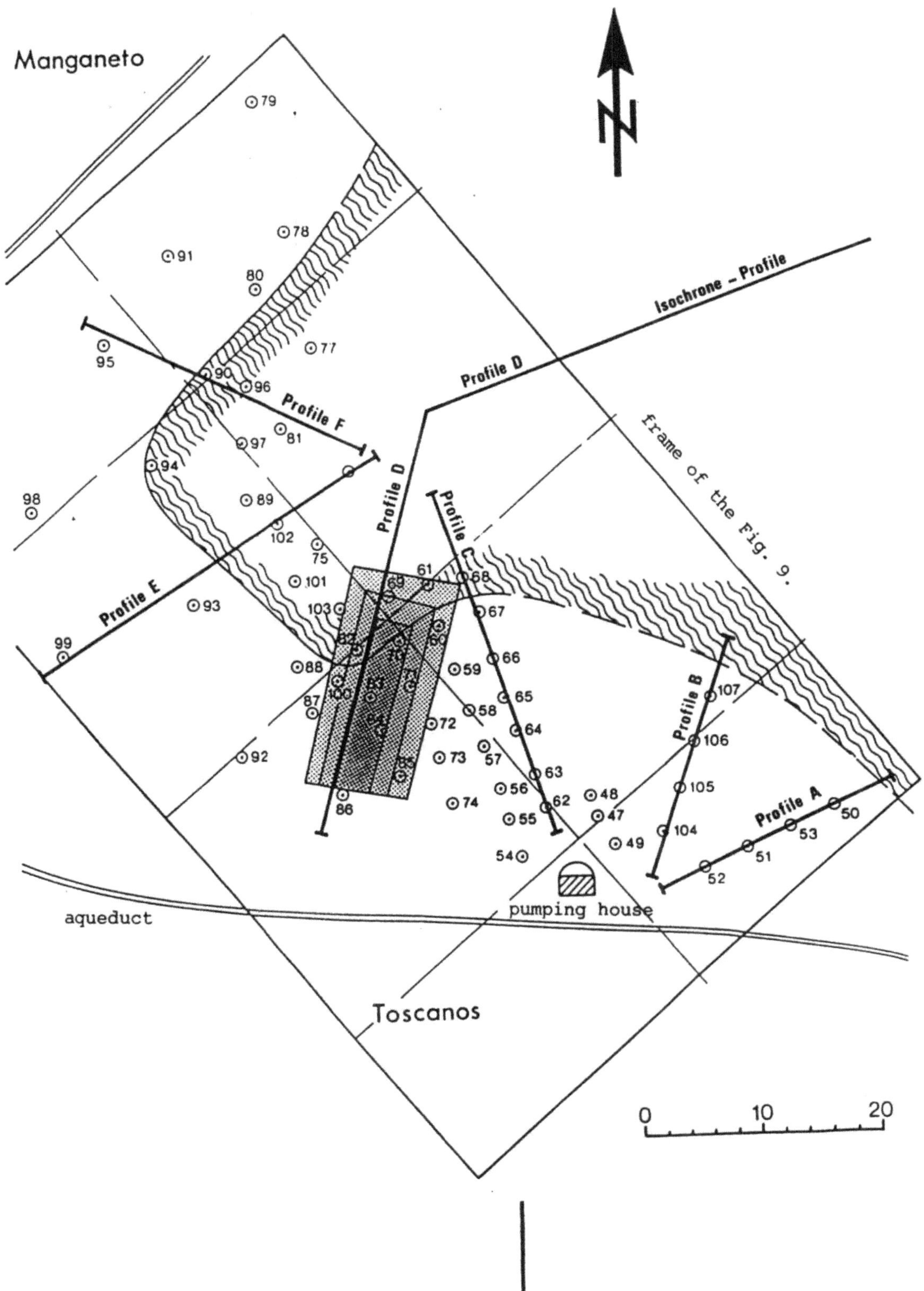

Figure 8. The "harbour bay" north of the settlement of Toscanos.

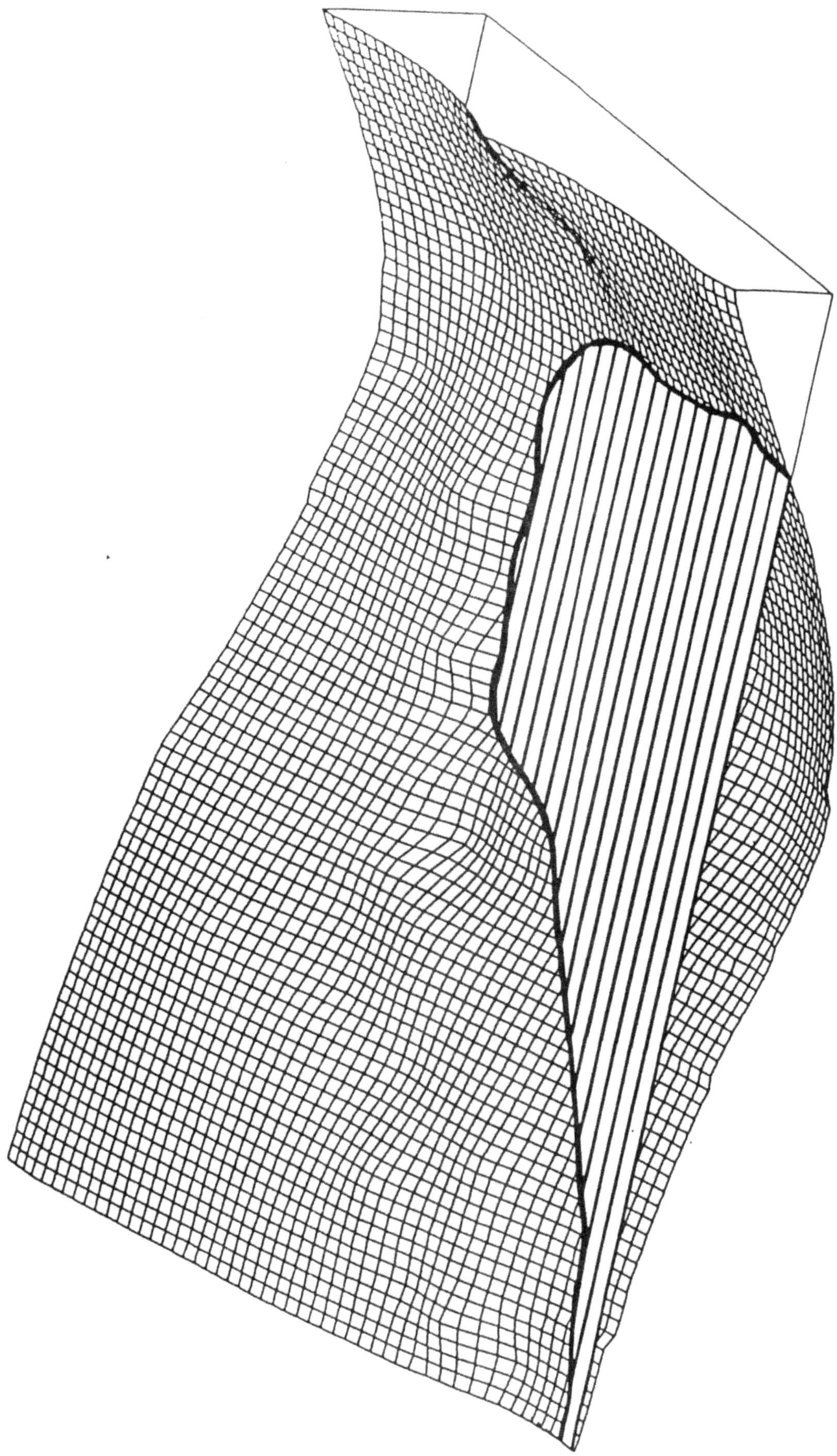

Figure 9. The surface of the "harbour bay" in the early Holocene until the 8th century.

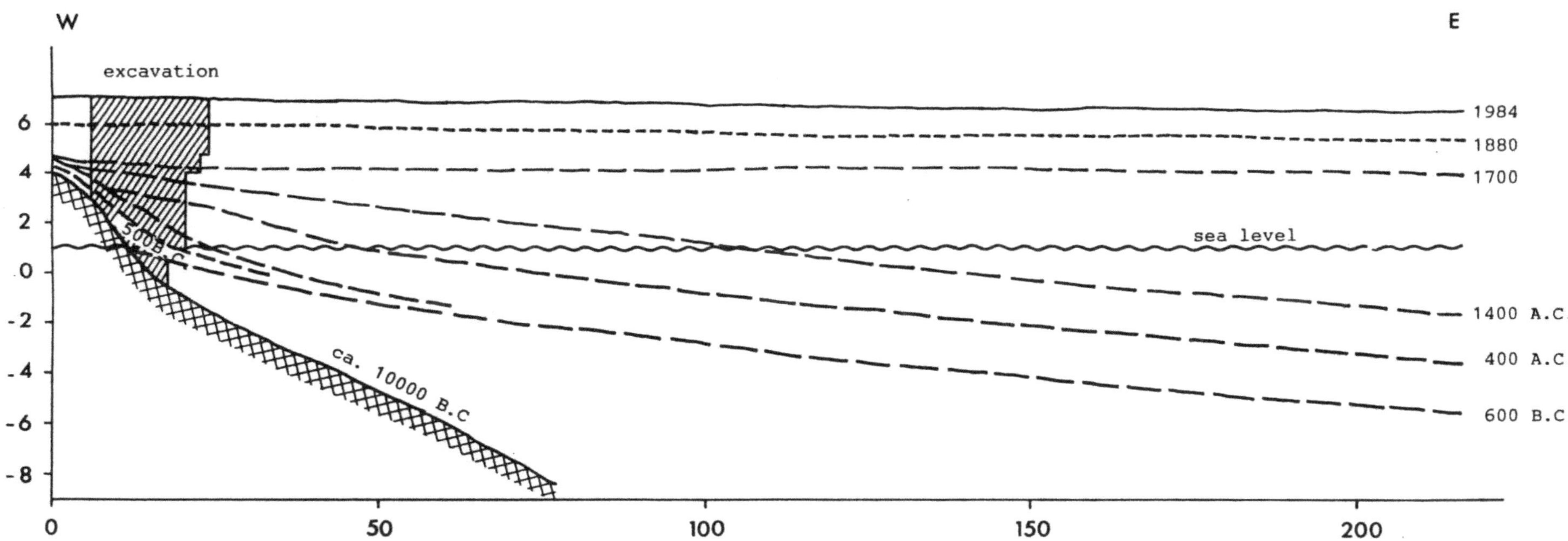

Figure 10. Isochronous profile.

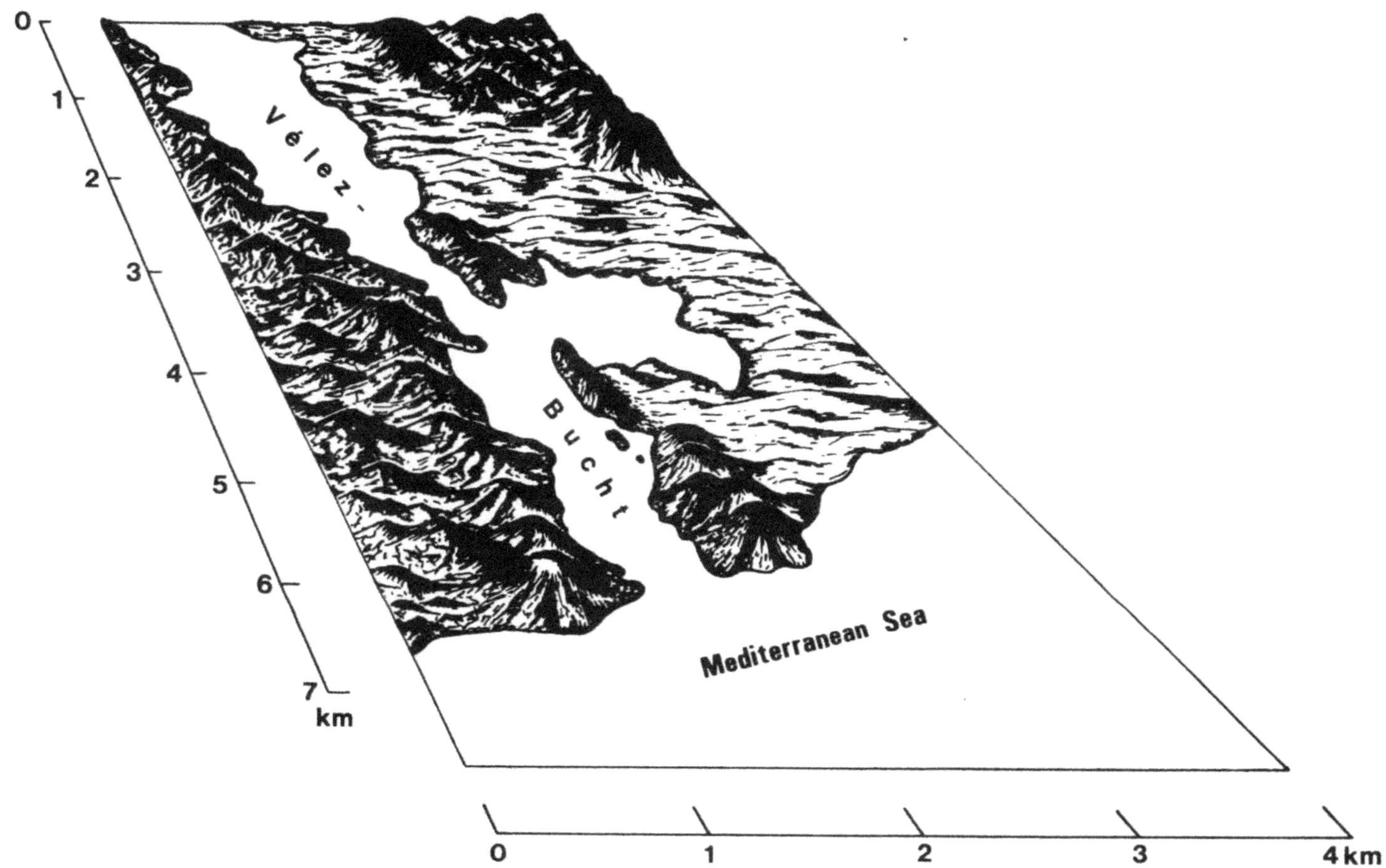

Figure 11. The estuary of the Rio de Velez after the Post Pleistocene rise of the sea-level and at the time of the first Phoenician settlements about 750 B.C.E.

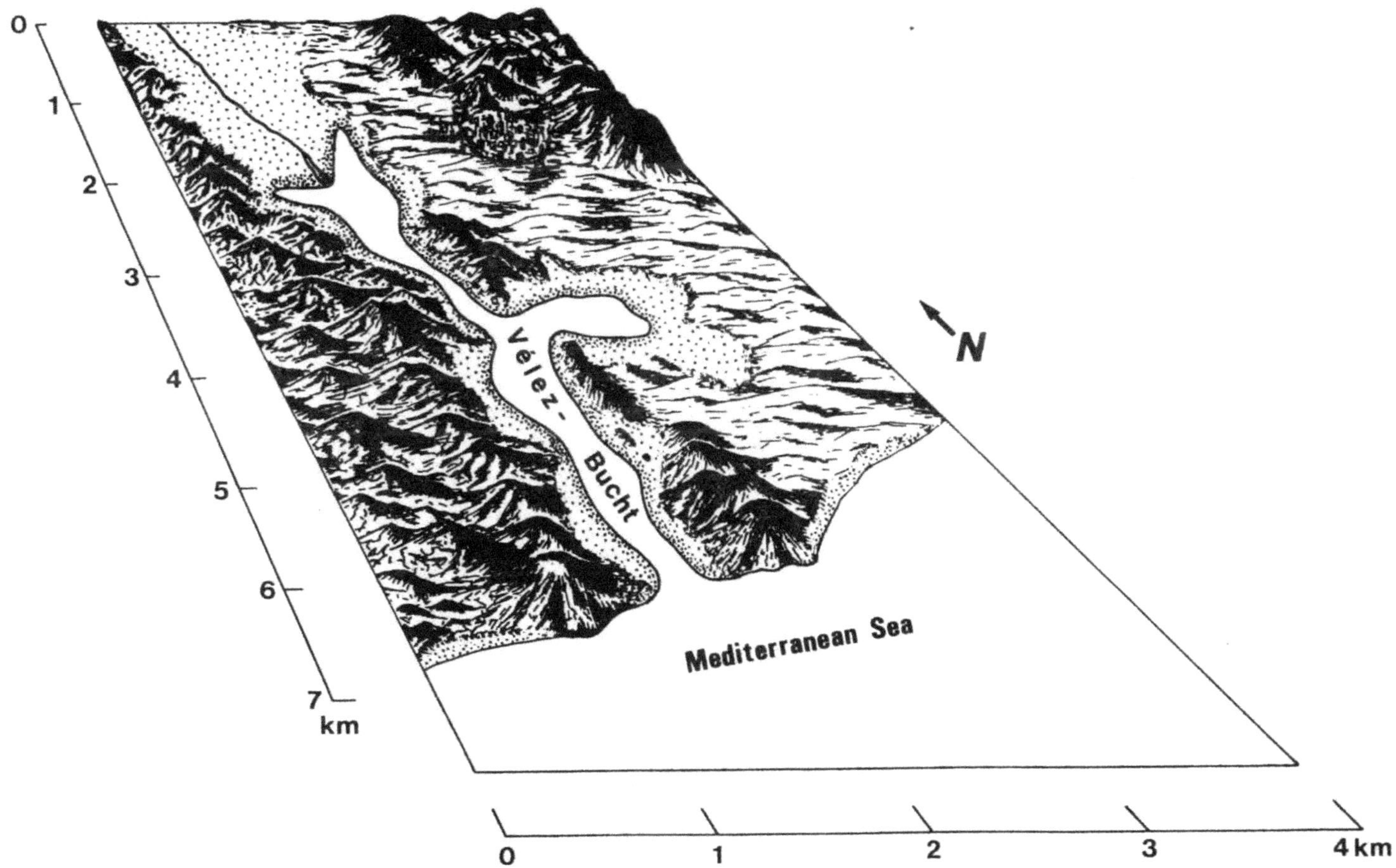

Figure 12. The estuary of the Rio de Velez in the 15th century, when the Catholic kings of Spain expelled the Moors from the southern part of the peninsula.

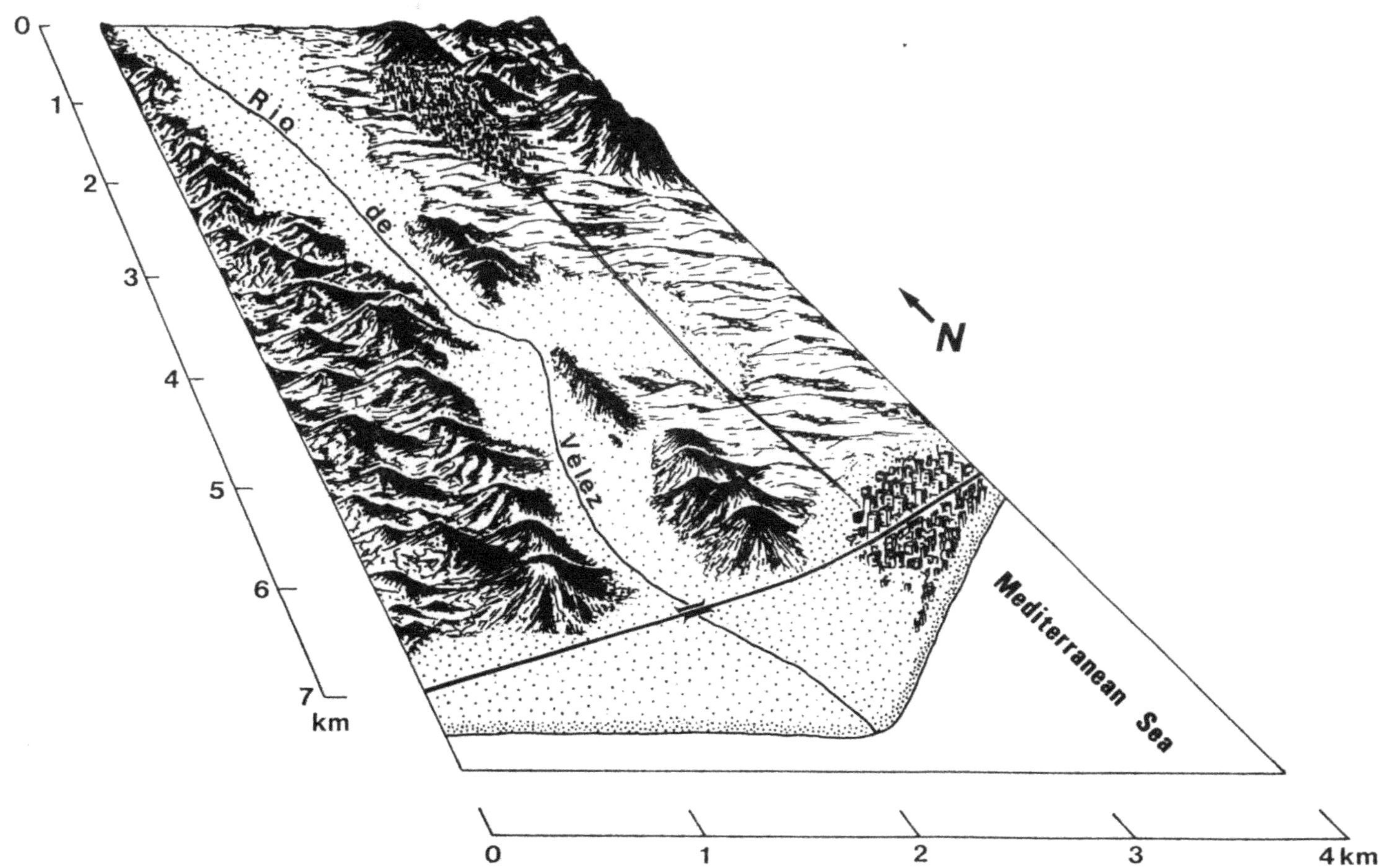

Figure 13. The current situation at the mouth of the Rio de Velez. Most of the young sediments were deposited during the 16th and 17th centuries.

EVANGELOS KAMBOUROGLOU, HAMPIK MAROUKIAN, ADAMANDIOS SAMPSON

Coastal Evolution and Archaeology North and South of Khalkis *(Euboea)* in the Last 5000 Years

ABSTRACT

The present study utilises the information gathered from 13 archaeological sites located along the coasts extending north and south of Khalkis in Central Greece in order to determine the changes that have taken place in the coastal environment. It also attempts to construct the sea-level curve of this area for the last 5000 years. It is concluded that there was a continuous submergence of all the sites studied until the Middle Byzantine times. The sea-level rise has been due primarily to eustatism and only to a limited degree to local tectonism. The sea-level curve agrees very well with others from other parts of the world. The connection of the northern and southern Euboic gulfs probably occurred at Khalkis 5-6000 years ago.

INTRODUCTION

The study of coastal changes is greatly facilitated by the presence of old settlements or structures along the coastline. Their vertical (elevation) or horizontal (distance) position in the coastal environment is essential in the determination of the extent of change that has taken place. Furthermore, the location of human occupation sites shows the significance that man attributed to certain parts of the coast for habitation, cultivation, leisure, commerce, industry and military purposes.

Along the extensive coastline of Greece, there are hundreds of ancient settlement sites which help in reconstructing the coastal environment of that time. The purpose of this study is to put together the information gathered from a number of such sites located north and south of Khalkis, between the island of Euboea and Boeotia in Central Greece (Fig. 1).

The city of Khalkis controls the Strait of Evripos joining the North and South Euboic gulfs (Fig. 2). It should be noted that the 20-metre isobath is only one kilometre north of Khalkis in the North Euboic gulf while the same isobath is about ten kilometres away in the South Euboic. Average tidal ranges are 42 cms. just north of Khalkis but only 9 cms. south of it (Zoi-Morou 1981, p. 72-75). A strong diurnal tidal current occurs in the narrow (a few tens of metres) Strait of Evripos.

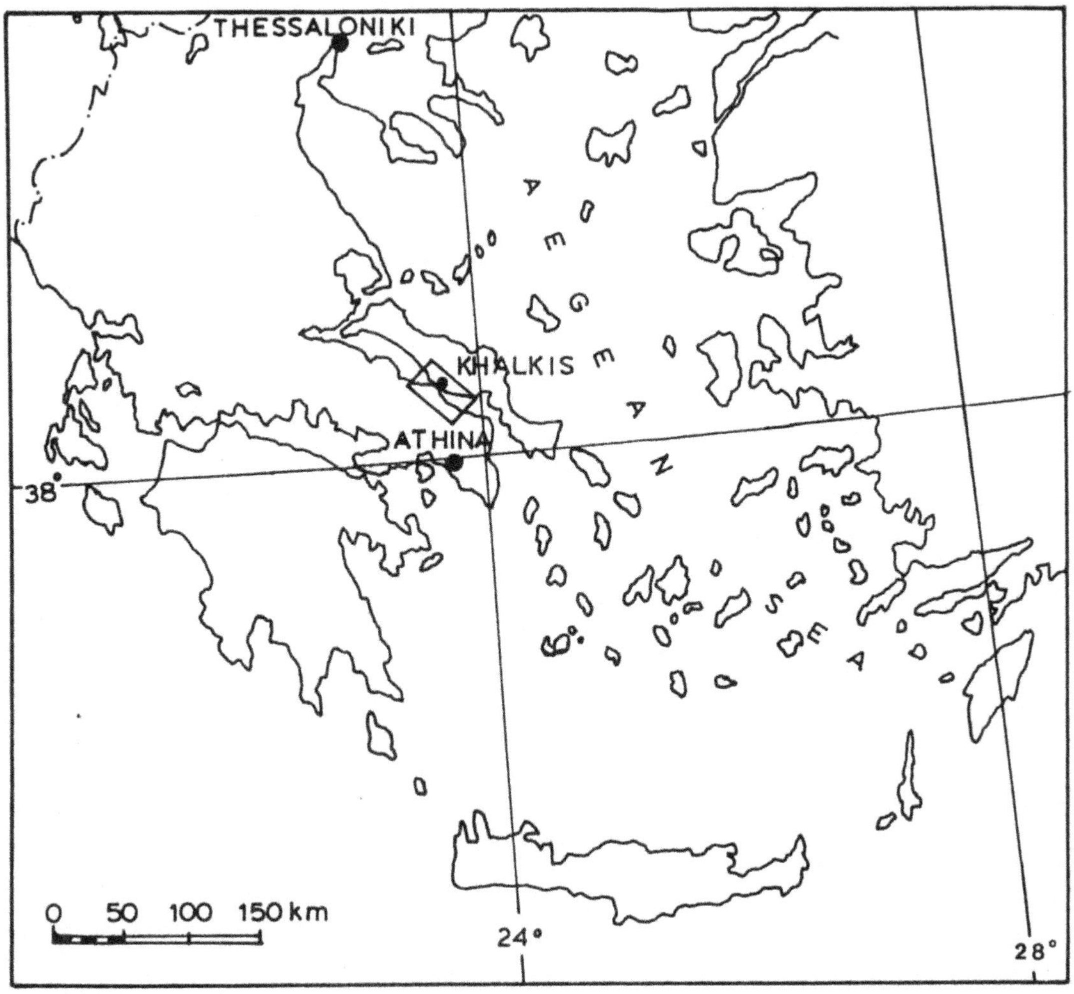

Figure 1: Location of study area around Khalkis.

The coastal environment of the area north and south of Khalkis is generally lowland. There is one important river named Lilas which flows into the South Euboic gulf. This river has formed a relatively small deltaic plain, less than twenty square kilometres in area.

The coasts from Anthidona to Salganeas have steeply dipping slopes down to the sea. Only the area around Anthidona has a gently rolling topography. The area surrounded by Salganeas, Bournondas, Khalkis and Glyfa is low, with a general dip of the land to the north-west. From Glyfa to Dilessi, we cross a relatively rugged mountainous topography, followed by a deltaic plain of a relatively small river (Remma Ritsionas) and just before Dilessi a few kilometres of eroding cliffs made of reddish unconsolidated sediments.

From Nea Artaki to Khalkis there is a coastal plain dipping gently to the sea but interrupted at places by high cliffs. Just south of Khalkis we encounter steeply sloping hills. From there on, we have the deltaic plain of the river Lilas and some small alluvial cones reaching down to the coast. Lefkandi is located on an eroding cliff believed to be a remnant of older alluvial deposits by Lilas. We should mention the presence of submerged beachrocks along many locations of the studied coasts. Some of these are found about a metre or more below today's sea-level, indicating positions of older sea-levels.

ARCHAEOLOGICAL SITES

The following thirteen archaeological sites were studied by the authors. Numerous studies made by others were also used. In some places, like Manika and Eretria, excavations are in progress. Others, like Lefkandi and Seimen, wait to be continued.

Anthidona

An ancient town located on the Boeotian coast, approximately 13 kilometres west-northwest of Khalkis. It is mentioned by Pausanias as being the last town of Boeotia. It prospered mainly in classical times, but its surficial ceramics show that there was some activity at least since the Geometric era. The same area was inhabited until Early Byzantine times (6th-7th centuries A.D.) to which the harbour structures belong (Georgiadis 1907, p.7). Their foundations belong to older periods dated at least to Roman times. It is considered to be one of the finest examples of ancient harbour constructions in the Aegean. According to Blackman (1973, p.124), a relatively small rise of sea-level since the construction of the harbour cannot be ruled out.

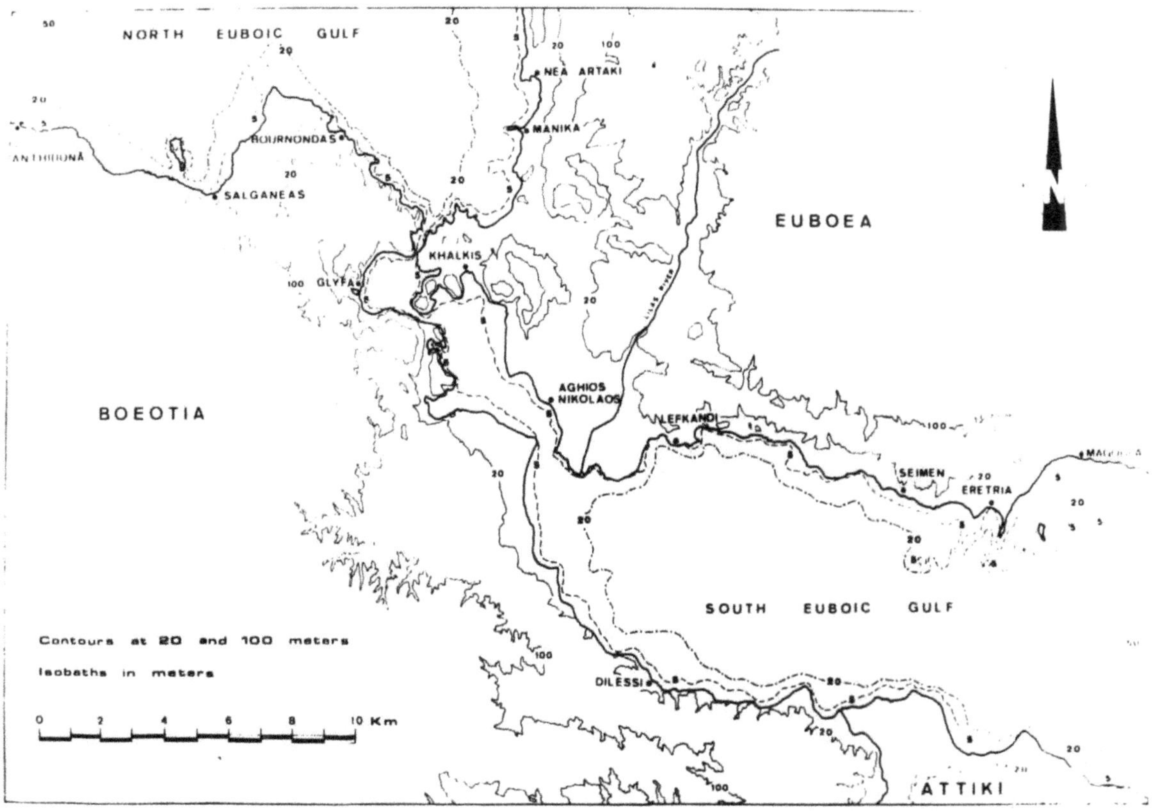

Figure 2: Coastal archaeological sites in the north and south Euboic gulfs mentioned in this study.

Salganeas

A prehistoric settlement on a low hill next to the sea. A preliminary excavation showed the stratigraphic column (Archaeol. Bull., 1970, p.212). The oldest settlement remains are dated to the Early Helladic II (2800-2300 B.C.E.) Life went on until the end of the Mycenaean age (1100 B.C.E.).

Bournondas

A prehistoric location with Early Helladic settlement remains on the coast and the island across. The strait between them is 1.5 metres deep and about 10 metres wide. It was probably joined to the land until recent times (Sampson 1985, p.358). The thin surficial deposits and the few Early Helladic remains (ceramics-obsidian) indicate that this area was sparsely populated. Most probably there was a small fishing settlement contemporaneous with that at Manika and similar to those found on other promontories of the Boeotian coast.

Figure 3: Late Byzantine and post-Byzantine remains (arrow) of buildings found at the head of the bay of Nea Artaki.

Glyfa

A Prehistoric site on a low hill contemporaneous with Salganeas. Founded in the Early Helladic II period, it was mainly inhabited during the Middle Helladic and Mycenaean times. An excavation conducted in 1976 along the southern flanks of the hill uncovered some buildings and a wall of Mycenaean age (13th cent. B.C.E.). The wall extends perpendicularly and very close to the coastline.

Dilessi

This coastal town is at present a summer resort. Near its centre, there are post-Classical and Hellenistic buildings (4th cent. B.C.E.), partially covered by the sea. They are related to the well-known temple of Delios Apollon. A hydrotherapeutic centre with spas was located in the same area (Bull. Corr. Hell. 1954, p. 132).

Nea Artaki

An Early Helladic site which has been eroded by the sea. Early Helladic sherds and obsidian were found here. The site is contemporaneous with Manika (2800-2300 B.C.E.). Late Byzantine and post-Byzantine remains of buildings are found at the head of the bay (Fig. 3). Some of these are covered by the rising sea (Sampson 1980, p. 358).

Manika

An extensive settlement and cemetery which prospered during the Early Helladic II and III (2900/2850-2350 B.C.E.) (Sampson 1985). A large part is covered by the sea. It was a real urban centre, the largest found in Greece so far. Remains of Early Helladic buildings are found down to a depth of 3.8 metres.

Khalkis (Aghios Stefanos)

Remains of large public buildings found during an excavation in 1984 whose foundations are at today's sea-level. They are dated to the 4th to 3rd centuries B.C.E. (Kambouroglou 1986).

Aghios Nikolaos

A late Roman coastal cemetery part of which is eroded by the sea, resulting in the exposure of tiled tombs at sea-level (Karapascalidou 1985, p. 14).

Lefkandi

A Prehistoric settlement of Early Helladic III (2450-2350 B.C.E.) age till the end of the Geometric period (700 B.C.E.) It prospered immediately after Manika. Following the end of the Geometric period, Lefkandi declined while Khalkis prospered (Popham and Sackett 1967, p. 60).

Seimen

A Neolithic coastal settlement dating from the Younger Neolithic II (4300-3800 B.C.E.). It was excavated in 1976 but the excavation stopped for technical reasons when it reached 70 cms. below present sea-level, while the archaeological bed was continuing (Sampson 1981, p. 129-131).

Eretria

Semi-submerged harbour structures of Early Archaic (?) times (6th cent. B.C.E.) Auberson and Schefold (1972, p. 29) acknowledge a sea-level rise of about 1.5 metres from the Early Archaic till today.

Magoula

A region which is characterized by continuous habitation from the Younger Neolithic till Roman times. Many tomb sites have been found. In particular, a coastal cemetery was found partly on land and partly submerged down to a depth of one metre (Kambouroglou et al. 1987).

DISCUSSION AND CONCLUSIONS

Following the analysis of the archaeological sites, it is evident that a rise of sea-level of at least 4.5 metres has occurred during the last 5000 years (Fig. 4). Manika, which is still being excavated and intensively studied both on land and underwater, has been precisely dated as have most other archaeological sites. The sites dating from the Archaic to Byzantine times are of particular interest and give a quite smooth sea-level curve.

The archaeological information leads us to the conclusion that the sea-level rise is due primarily to eustatism and to a very limited extent to local tectonism. The sea-level curve agrees very well with those from other parts of the world (Shepard 1963; Jelgersma 1966; Bloom 1967; 1979; Mörner 1971). The general retreat of the coasts of the area studied is one more indication for advocating a sea-level rise. This retreat seems to be continuing today due also to the intervention of man in the coastal environment and the ensuing disturbance of certain processes which operate on the coasts.

Finally, we believe that the connection of the North and South Euboic gulfs probably took place 5-6000 years ago. The newly created marine conditions at the Strait of Evripos at Khalkis should have changed radically, such as the phenomenon of the strong tidal currents observed there today.

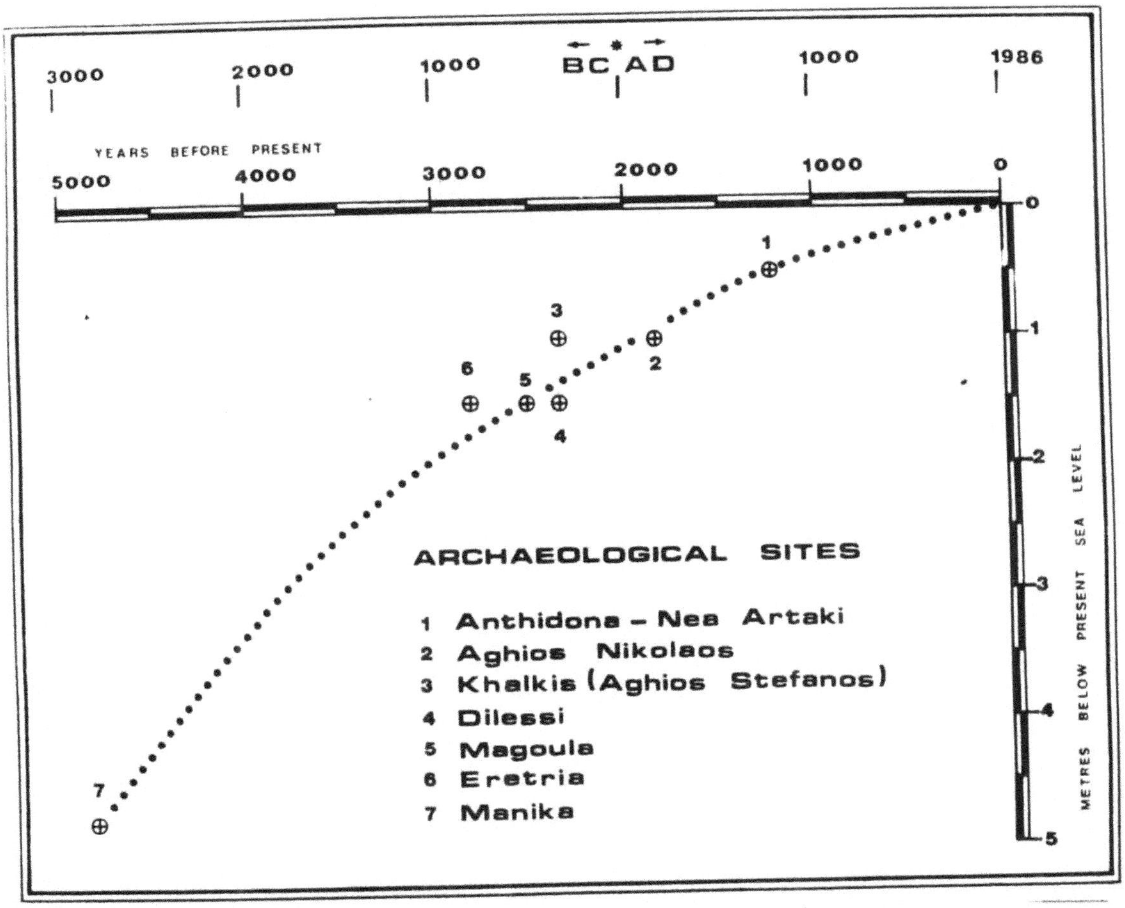

Figure 4: Sea-level curve based on information gathered from Archaeological sites found in the area of Khalkis.

REFERENCES

Auberson, P. and K. Schefold. 1972. Führer durch Eretria. A. Francke AG Verlag, Bern.

Bloom, A.L. 1967. Pleistocene shorelines: a new test of isostasy. Bull. Geol. Soc. America, 78:422-436.

Bloom, A.L. 1979. Atlas of Sea-Level Curves. Cornell University, New York.

Blackman, D.J. 1973. Evidence of sea-level change in ancient harbours and coastal installations. In: Marine Archaeology. D.J. Blackman (Ed.), Colston Papers 23:115-139.

Georgiades, A. 1907. Les Ports de la Grèce dans l'Antiquité qui Subsistent Encore Aujourd'hui. Athens.

Jelgersma, S. 1966. Sea level changes during the last 10000 years. In: World Climate 8000 to 0 B.C., J.S. Sawyer (Ed.), Proc. Intern. Symposium on World Climates, 18-19 April, 1966, Royal Meteorological Society, London.

Kambouroglou, E., A. Karapascalidou and S. Manolis. 1987. An Archaeological Coastal Cemetery in the Area of Eretria (Euboea) and Its Contribution to the Study of Its Geomorphological Evolution. Proc. Anthropological Symposium 28-30 November, 1986, Greek Anthropol. Society, Athens, (in Greek).

Kambouroglou, E., A. Sampson and H. Maroukian. 1986. The Coastal Topography and Archaeology of Manika: an Early Helladic town in east Central Euboea, Greece. Abstracts, 25th Intern. Symp. on Archaeometry, 19-23 May, 1986, Athens.

Karapascalidou, A. 1985. A Roman settlement in the vicinity of Chalkis. Old World Archaeology Newsletter, IX(3):14.

Mörner, N.A. 1971. Eustatic changes during the last 20000 years and a method of separating the isostatic and eustatic factors in an up-lifted area. Palaeogeogr. Palaeoclimat. Palaeoecology, 9:153-181.

Popham, M. and L. Sackett. 1967. Excavations at Lefkandi, Euboea, 1964-1966, London.

Sackett, L., R. Hankey, J. Howell, T. Jacobsen and N. Popham. 1967. Pre-historic Euboea. British School of Archaeology, Athens.

Sampson, A. 1980. Prehistoric sites and settlements in Euboea. Archive of Euboic Studies (Αρχείο Ευβοϊκών Μελετών) 91-249 (in Greek).

Sampson, A. 1981. The Neolithic and Early Helladic I in Euboea. Doctoral dissertation, Athens (in Greek).

Sampson, A. 1985. Manika, an Early Helladic Town in Chalkis. Athens (in Greek with extensive English summary).

Shepard, F.P. 1963. Submarine Geology. New York (2nd edition).

Zoi-Morou, A. 1981. Tidal Data of Greek Ports. Greek Hydrographic Service, Oceanographic Study no. 13, Athens (in Greek).

JOHN LORET

Salvage Archaeology Beneath the Streets of New York City

ABSTRACT

Much of the existing shoreline of Manhattan Island is "man-made" land. The landfill process, which started with the early Dutch settlers as "poldering", continues even now as in the 92-acre Battery City project. From the seventeenth to the early nineteenth century, land-filling extended the original shoreline of the East River from Pearl Street through Water Street, then to Front Street, and finally to South Street. On the West side, land-filling of the Hudson River started at Greenwich Street, to West Street. Consequently, much of the remains of the early port of New York lies beneath the streets of Manhattan.

The pre-revolutionary port of New York was a hub in the transatlantic trade network for eighteenth century "Indiamen" merchant ships. The wharves and basins became depositories for many cultural items which were lost or inadvertently dropped overboard during cargo on-and off-loading.

The construction of modern skyscrapers along shoreline areas require that foundations reach solid bedrock. This necessitates the excavation of old landfill areas, unearthing entire ships, industrial structures and artifacts representing over 300 years of history.

INTRODUCTION

The event that opened the port of New York started with the dropping of the anchor of the "Half Moon" by the explorer Henry Hudson in what is now lower New York Harbour, in September 1609. Since its beginning as a Dutch East India Company trading post, the port of New York has depended upon the sea for shipping goods for profit around the world.

The destiny and success of this seaport city was secure and grew mainly because of its ideal geography. The New York port could be entered from two directions: directly from the Atlantic Ocean from the south and east and from Long Island Sound into the east or Sound River from the north. The inner, or upper, harbour is well protected from the gales, hurricanes and heavy seas of the Atlantic Ocean by Long Island in the east and Staten Island to the south. In the cold months, the port remains ice-free. This was an especially crucial factor in the days when ships were constructed of wood. Also, the New York port has immediate access to major inland waterways for trade routes to the interior of the continent. (Shumway, 1975).

THE LANDFILL PROCESS

In the seventeenth century, the small Dutch trading post grew into a complex town. In 1664, the English took over the city, then called New Amsterdam, and renamed it New York.

By the end of the seventeenth century, the business section was becoming overcrowded and city officials decided to extend the land areas by filling along the East and Hudson Rivers. This activity, still continuing today, (i.e. the 92-acre "Battery City" project), began with the early Dutch settlers and was called "poldering". (Greenhouse, 1983). From the seventeenth century to the early nineteenth century, landfilling continued and extended the original shoreline of southern New York City to South Street on the East River. On the Hudson River, the filling started at Greenwich Street to West Street (the West Side Highway). (Figure 1).

City officials encouraged this process to boost commerce, trade and increase settlement, which would add to the tax rolls. Filling in low-lying areas was also seen as a means of reducing sickness and epidemics. Yellow fever was particularly rampant in the eighteenth century in New York. The Montgomerie Charter, extended by the English crown in 1873 to New York, granted the city the right to underwater land up to 120 metres beyond low-water. This led to favouritism given in underwater purchase privileges to government officials, well-known citizens, and prominent businessmen. (Peterson, 1917).

Merchants purchased these parcels of liquid real estate inexpensively and filled them at their own expense with trash and other unwanted materials, i.e. bricks, old vessels, stones, earth, broken pottery, china, glassware, etc. Consequently, many of these artifacts and much of the remains of the old port of New York lie today beneath the streets of lower Manhattan. (Kearns, 1986).

The construction of modern skyscrapers, such as the World Trade Center, along current and near by shoreline sites, necessitate the excavation of old landfill areas. In these excavations, old ships, wharves, early industrial structures, trade tools and artifacts representing 300 years of history of the old port of New York are being uncovered and revealing many aspects of the city's early growth and development.

Until recently, new city environmental review processes administered by the Landmarks Preservation Commission were not enforced. Historic sites were destroyed by the contractor's equipment and carted off to dump sites. The only opportunity archaeologists and historians had to study the sites or obtain artifacts, was to attempt to enter construction sites during the off-work hours of the construction team or to follow the trucks to the dump site and sort through the demolition debris. Unfortunately, in past years, many important sites were completely disassembled, dumped and covered before any professional archaeologist or historian could be available even to photograph an excavation.

Figure 1: Map of lower Manhattan Island. Shaded areas represent landfill. The area not shaded is the original shape of the Island as settled by the Dutch. Numbers 1-4 indicate the archaeological site mentioned in this paper. (from Townsend 1909).

Although some early eighteenth and nineteenth century city records document many of the changes and developments in the old port of New York, the details and specifics about engineering techniques, materials and technologies used are not given. This information can only be obtained by the study and analysis of the structures and the artifacts themselves.

Since 1978, through the efforts of the City Environmental Quality Review (CEQR), eight sites have been properly investigated. (Geismar, 1986).

THE EARLY DUTCH SHIP "TIJGER"

In 1611, the adventurous Adriaen Block, a Dutch lawyer, gave up his practice to follow the sea as a fur-trading merchant. He, and a partner, Hendrick Christioensen, chartered a vessel and sailed from Amsterdam to the Hudson River in the New World. The enterprise proved successful and they returned to Holland with an excellent cargo of valuable pelts. This stimulated other Dutch merchants who believed the fur-trading venture could be highly profitable.

On a third expedition in November 1613, Adriaen Block's ship "Tijger" (Tiger), rode at anchor in the Hudson River, deeply laden with a cargo of beaver, otter and other pelts. Fire broke out on board and soon raged out of control. Block and his crew salvaged a few sails, line, tools and fittings, placing them into the ship's tender, and escaped from the flaming vessel. The "Tijger" burned to the water line, her cargo of pelts a total loss, leaving only a charred hull, which was later beached in a little cove on the east bank of the Hudson River near what is now the west side of Greenwich Street near Dey Street. In a map developed by the Underpinning and Foundation Company Inc., showing the geography of Manhattan Island from early Dutch times, a stream is shown at this location. (Townsend, 1909).

With the assistance of friendly Indians, Block and his crew set to work laying the keel of a smaller new ship. In the spring of 1614, they launched the ship which Block named "Onrust", meaning restless. Block thus became New York's first shipbuilder and the "Onrust", the first ship to be built there. Block sailed the "Onrust" north in the East River, through Hell Gate, and discovered Long Island Sound. He continued to explore the coastal areas of Connecticut, Rhode Island and Massachusetts. He named Block Island after himself and charted the coastline as far north as Pye Bay, north of present day Boston. (Williamson, 1959).

In 1916, 303 years after Adriaen Block beached the hull of the "Tijger", a three metre section of charred keelson and three frames were unearthed during the excavation for the Interborough Rapid Transit Company (I.R.T.) subway at the intersection of Greenwich and Dey Streets. The section of the hull believed to be from the "Tijger" was uncovered at a depth of six metres below street level, along with a Dutch broad axe, trade beads, clay pipes, chain, cannon balls and

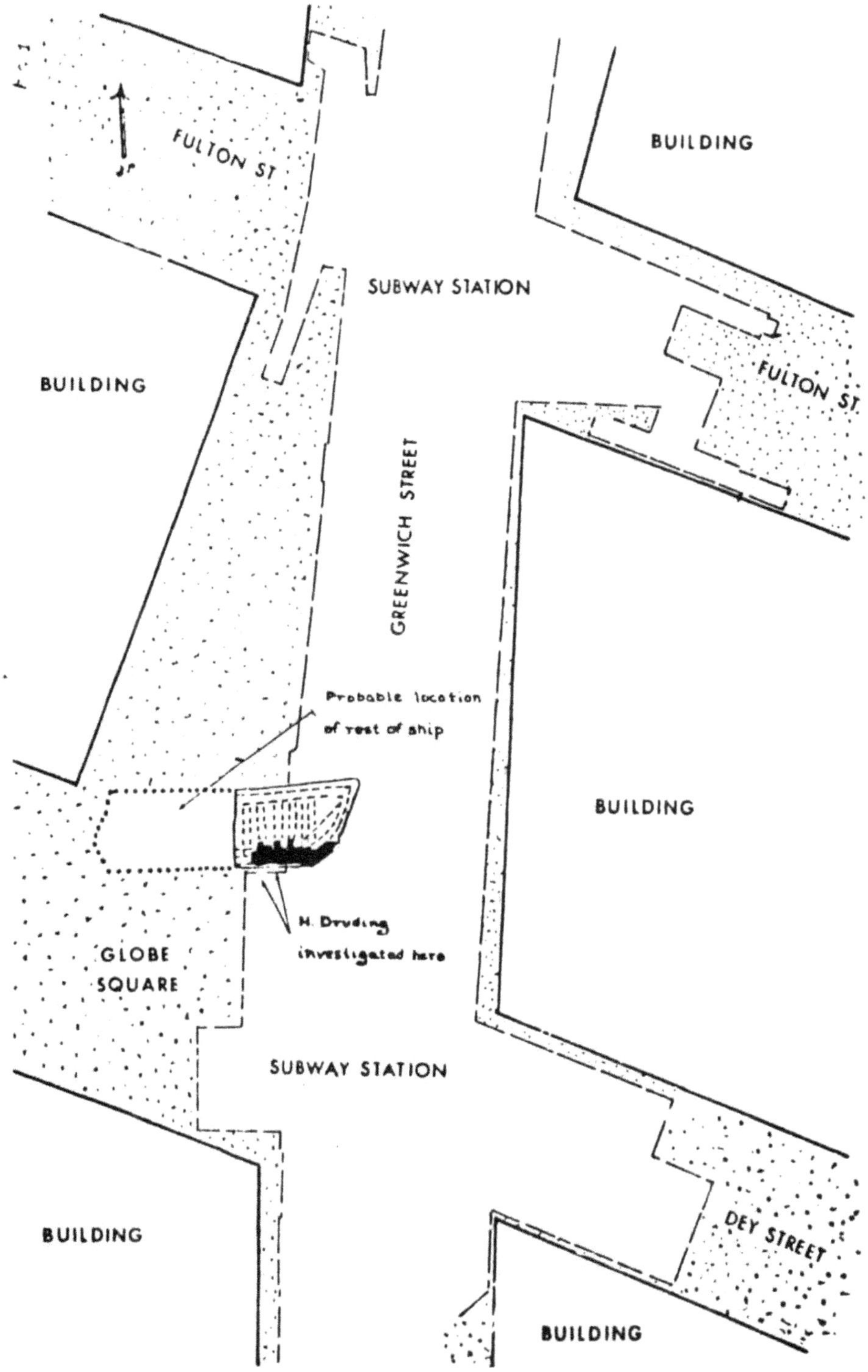

Figure 2: Copy of Board of Transportation map showing the position of the section of bow believed to be the "Tijger" when discovered in excavating the subway in 1916. The site is now within the heart of the World Trade Center.

pottery. The top 2.5 metres of material consisted of landfill that had been deposited during shoreline extension 200 years earlier. The lower 3.5 metres consisted of river sediments. (Solecki, 1974). The entire discovery would have been destroyed had it not been for the efforts of James A. Kelly, the job foreman, who later became the historian for

Kings County. Wood from the keel was radio-carbon dated to be 420 years old ±80, L-262. (Broecker, 1957). The Research Laboratory of Bethlehem Steel Company, Pennsylvania, conducted composition tests of an iron bolt taken from the hull. The results indicated that the iron used had been forged about the year 1600. (Solecki, 1974). (Figure 2).

Kelly attempted to pull the remaining section of the ship out of the excavation with mules, but was unsuccessful. Therefore, the exposed 3 metre section was cut off, leaving at least 15 metres more of the "Tijger" buried west of the subway excavation. The recovered hull section is currently on display at the Museum of the City of New York (Figure 3).

With the construction of the World Trade Center in the 1960's, adjacent to the 1916 subway excavation site, another attempt was made to recover the remaining section of "Tijger" hull under the direction of Dr. Ralph Solecki, Professor of Anthropology, Columbia University. After weeks of work, no sign of the ship was found. However, three of the six bronze breech-loading guns were recovered. The removable breech of the guns contain the engraved monogram "VOC", for Vereenigde Oostindische Companie, the Dutch East India Company. (Figure 4).

EIGHTEENTH CENTURY TANNERY

At the corner of Water and Fulton Streets (199 Water Street) (Figure 1), the remains of a small industrial complex believed to be an eighteenth century tannery was uncovered in 1981. Several cisterns and pits, essential to the tanning technology of the period, were uncovered. (Figures 5 & 6). The tanning of leather was a messy process requiring access to a large amount of flowing water. After the animals were slaughtered and skinned, the hides were washed in the river to remove grime, blood and loose bits of flesh attached to the skin. So urban riverside properties were essential for the tanning process. In pre-industrial times, the river area was more often what the railroad districts later would become in the nineteenth century. After cleaning in the river, hides were placed in cisterns for additional soaking in a lime bath solution (Figure 7). Liming loosened the hair and began the curing process. After soaking in lime for a week, the hides were removed and stacked for two weeks. Skins were then again placed in lime - this process would continue for four to six months.

After liming, the hides were placed in tanning pits with tree bark containing tannic acid. The bark from oak, willow, chestnut, hemlock, mimosa and other available species could be used. Hides were laid out in the pit on a layer of bark, then covered with bark and stacked alternately until the cistern or pit was filled and capped with a thick layer of bark. Hides would remain in the pit for several months. When unpacked, the hides would be soft and pliant. The entire process would take up to a year. (Hunt, 1867).

Figure 3: Timbers of the section of the ship believed to be Adriaen Block's "Tijger" burned off lower Manhattan in 1613; the section was recovered from the subway excavation in 1916 (Museum of the City of New York).

EARLY EIGHTEENTH CENTURY MERCHANT SHIP

About 1740, dock workers used a merchant ship as part of land-fill extending a pier out into the East River. In the winter of 1981, archaeologists investigating a pre-construction site in a parking lot at 175 Water Street (Figure 1) discovered a 250-year old merchant ship buried eight feet below street level. The owner and developer, Howard R. Ronson, had purchased the land to build a 30-storey office building, but unknowingly also acquired the buried ship. The unknown ship was, therefore, referred to as the "Ronson Ship". Mr. Ronson gave the archaeologists one month and financed the professional excavation of the site. The delay in construction cost Ronson an additional $20,000 per day.

Except for a few old prints and drawings, no one knew what an early eighteenth century merchant ship looked like, what materials were used, or how she was fastened. The only other known ship of this period is the wreck of the "Amsterdam", sunk off England, which has limited access, even to divers, due to extremely harsh tidal conditions. (Dean, 1985).

In one month, the 35-member excavation team measured, sketched, photographed and recorded almost every piece of the ship. 310,000 artifacts were recorded and over 17 tons of fill were removed. (Geiser, 1986). The cost to remove and conserve the entire ship would have been prohibitive. However, Mr. Ronson offered to underwrite the conservation of the ship's unique bow. The bow section was disassembled, carefully packed and transported to a conservation laboratory in Groton, Massachusetts.

Figure 4: One of the three bronze breech-loading guns recovered from the excavation for the World Trade Center in 1968. These guns are believed to be from the "Tijger". The removable breech contains the monogram "VOC", for Vereenigde Oostindische Companie (Dutch East India Company). The wooden carriage was not recovered with the gun, but constructed for exhibit. (Science Museum of Long Island).

The ship was larger than the average New York merchantman, being 170 to 210 tons. She measured 24.6 metres between perpendiculars and 21.5 metres on the keel. The maximum beam was 7.8 metres and the draft 3 metres. A good deal of information about early eighteenth century ship construction was obtained. The bow was almost squared or "apple cheeked"; a similar form is shown in a 1717 engraving of New York harbour. (Figure 8). Unfortunately, nothing in the excavation was found to indicate where she was built. However, the U.S. Forest Service Laboratory in Madison, Wisconsin, has analyzed wood samples taken from the ship. Her deck is long leaf pine Pinus australis, caulked with pine pitch, sawdust and hair. Her tree nail fastenings are hickory Carya sp. or juniper Juniperus sp., with some iron. The hull planks and frames are white oak Quercus alba. Wood analysis, however, may not settle where the ship was built, since the British shipyards in this period imported American timber. The problem is further complicated in that many wood analysis specialists claim that some woods, i.e. American white oak Quercus alba and European white oak Quercus ruber are impossible to tell apart. (Riess, 1985).

The bow timbers have since been moved from the Groton, Massachusetts laboratory to the Mariners Museum in Newport News, Virginia, where Warren Riess is developing a major permanent exhibition on the "Ronson Ship".

Figures 5 and 6: This 18th century cistern and pit complex was uncovered at 199 Water Street. They probably indicate the site of a Tannery. Tannery technology for the period required access to large amounts of water. This complex was on the shore of the East River at the time.

Figure 7: An 18th century engraving showing the soaking of hides in cisterns. Soaking was done in a solution of lime to loosen hairs to be removed before the tanning process. (Diderot, 1959).

THE ASSAY OFFICE SITE

An example of another site from which there is documented historical evidence from the period in which the underwater property was purchased, then filled, built upon and then excavated, is the Assay Office Site, also known as Block 35. (Figure 1).

In 1775, the Assay Office Site was a large water-filled basin, part of the East (Sound) River with a water depth of 5 to 7.5 metres. (Ratzer, 1776). On March 21, 1775, the City granted water rights of this site to several individuals who were known merchants and politicians. Landfilling was started shortly thereafter. (Ewen, 1827-30). It was not until an assessment was taken of Front Street in 1790, that we have the first specific evidence of the nature of the landfill. This document also identifies the function of the properties located on the filled block for this period. (Figure 9).

The map shows that the site had three wharves extending into the East River with the above named properties located on landfill extending eastwards halfway across the site from Front Street. The east ends of the three wharves, Gouverneur's, Bache's and Roosevelt's, would later become the beginning of South Street. It is believed that the fill to this point was completed between 1768 and 1790. (Greenhouse, 1983).

FILLING IN TO SOUTH STREET

In 1795, the Common Council became concerned with the "unlimited extension" of water-lot development, considering the activity "injurious, if not ruinous", because of drainage and health problems. The Council ordered that a 21 metre wide street (modern South Street) be laid out as an "outer street" as an "ornament to the City", beyond which no additional grants were to be made and no buildings or wharves constructed, except for those "deemed necessary for the safety of Shipping and the convenience of Trade and Navigation". (MCC, II, 420-421).

In October 1797, the Common Council ordered that all proprietors of water lots have their lots extended and filled to South Street with "wholesome" or sterile fill to a "proper height" by October 1, 1798. It was not until 1804 that the filling operation was completed. (CPW, October 8, 1804).

DEVELOPMENT: NINETEENTH CENTURY

New York City continued to rise as a shipping and mercantile centre and it soon became the leading port in the country. In the early part of the nineteenth century, conditions in the South Street section of New York were crowded, confused and noisy. Commerce expanded and dock space became more valuable. Also the site and draught of larger ships, such

Figure 8: The "apple cheeked" bow section of an early 18th century merchantman uncovered at 175 Water Street. The unknown ship is referred to as the "Ronson Ship" after the name of the owner and developer of the property. (Riess, 1985).

as the clippers and steam ships, outstripped the older eighteenth century wharves. (Greenhouse, 1983).

On December 16, 1835, a fire spread through the entire area destroying over 170 buildings, including the Assay Office site. While the fire, later known as the "Great Fire of 1835", did much to change the physical appearance of Block 35, and the adjacent environs, it did not alter the character of the area. (Foster). South Street was still a scene of busy life and commerce. Harbour masters, stevedores, longshoremen, sailors and ship owners mingled with commission merchants, grocers and chandlers.

In spring 1983, excavation began on the Assay Office site. Uncovered were a complex of structures of eighteenth century wharves, in an excellent state of preservation. (Figure 10). This discovery provided an unusual opportunity to study the techniques used by the builders of these crib-like structures of timber and stone. The wharf construction resembles those found in fill at ancient Roman and Mediaeval European port sites. In fact, iron-tipped pilings recovered after the wharves were disassembled, show a technology that dates back to the Roman times in bridge construction. Similar pilings have been recovered in old Roman bridges in Italy. The bridges across the Moselle River in Trier,

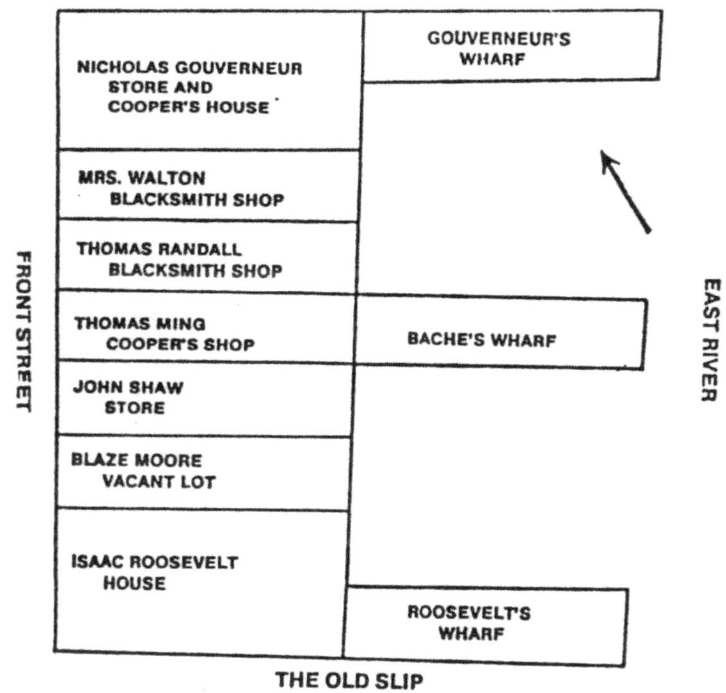

Figure 9: The Assay Office site reconstructed from 1790 tax lists. (Greenhouse, 1983).

Figure 10: This heavy timber wharf complex in an excellent state of preservation was uncovered in excavation of the Assay Office site in 1983. The builders used crib and cobble construction; this construction resembles that which was used in early mediaeval European port sites.

Germany, and old sections of the London Bridge in England, were also constructed by the Romans and have iron-tipped pilings. (Loret, 1986). (Figure 11).

The Roman wharves were built prior to the deforestation of the European forests. The use of heavy timbers in the eighteenth century lower Manhattan wharves were also due to the availability of large suitable native trees of the New World's primaeval forests.

After the wharves were examined and photographed, the contractor began to disassemble them. Through the efforts and persistence of an amateur archaeologist and historian, William Asadorian, who followed the daily work of the construction crew, a remarkable assemblage of artifacts was discovered and salvaged. The collection was uncovered within the silt and clay sediments of the river bed at a depth of 10.5 metres below street level. Items included the remains of two distinctly different small craft, an assortment of blacksmith's and shipwright's tools, an oil lamp, navigation dividers, hundreds of shards of porcelain tiles, china, glass, clay pipes, and stoneware. Also found were an assortment of organic artifacts including textiles, i.e. gloves, felt hats, rattan, tar brushes and rope; leather objects including shoes,horse collar, chafing gear and bits of leather clothing; wooden objects, i.e. rigging blocks, sheaves and a fid used to splice line. These objects obviously represent several maritime crafts and trades associated with the eighteenth century and will be of value in reconstructing conditions of the New York Port for that period.

CHINESE EXPORT PORCELAIN

The fragments of Chinese porcelain found represent what is called "Chinese Export Porcelain", and are examples of "common wares" or "blue-and-white" wares that were produced in China especially for the foreign trade, and specifically for the United States market.

Trade between China and Europe began in the middle of the sixteenth century. Chinese-American trade, however, did not begin until after the American Revolution when the new nation had sufficient ships and manpower to compete in the foreign trade. The first American vessel to engage in this trade was the "Empress of China", which first arrived at Macao harbour in August 1784, and returned to New York in May 1785. This does not necessarily mean that these fragments are post-1785. Shards of blue-and-white ware have been found in Florida, Georgia and Virginia in sites which were occupied prior to 1785. Therefore, we conclude that some Chinese blue-and-white ware had reached America via Europe prior to the American Revolution and the establishment of the American-Chinese trade. (Figure 12).

Once established, trade with China flourished until about the middle of the nineteenth century. In the 1840's additional ports were opened to traders, producing alternating periods of over-supply and shortages. Mass production in large quantities to meet the demand resulted in

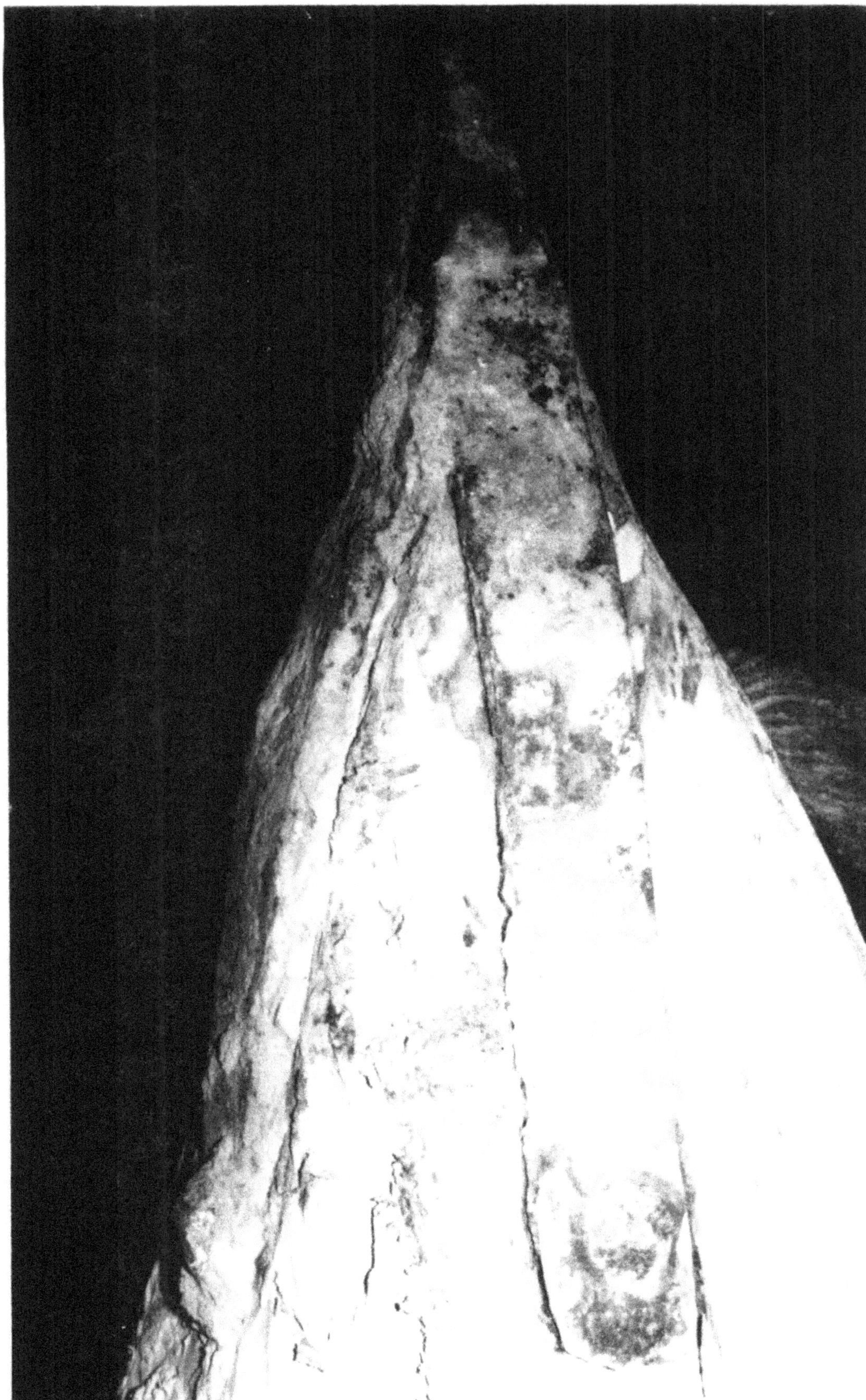

Figure 11: One of the iron-tipped pilings recovered from the wharves of the Assay Office site. Similar technology was used by the Romans in bridge construction. (Loret, 1986).

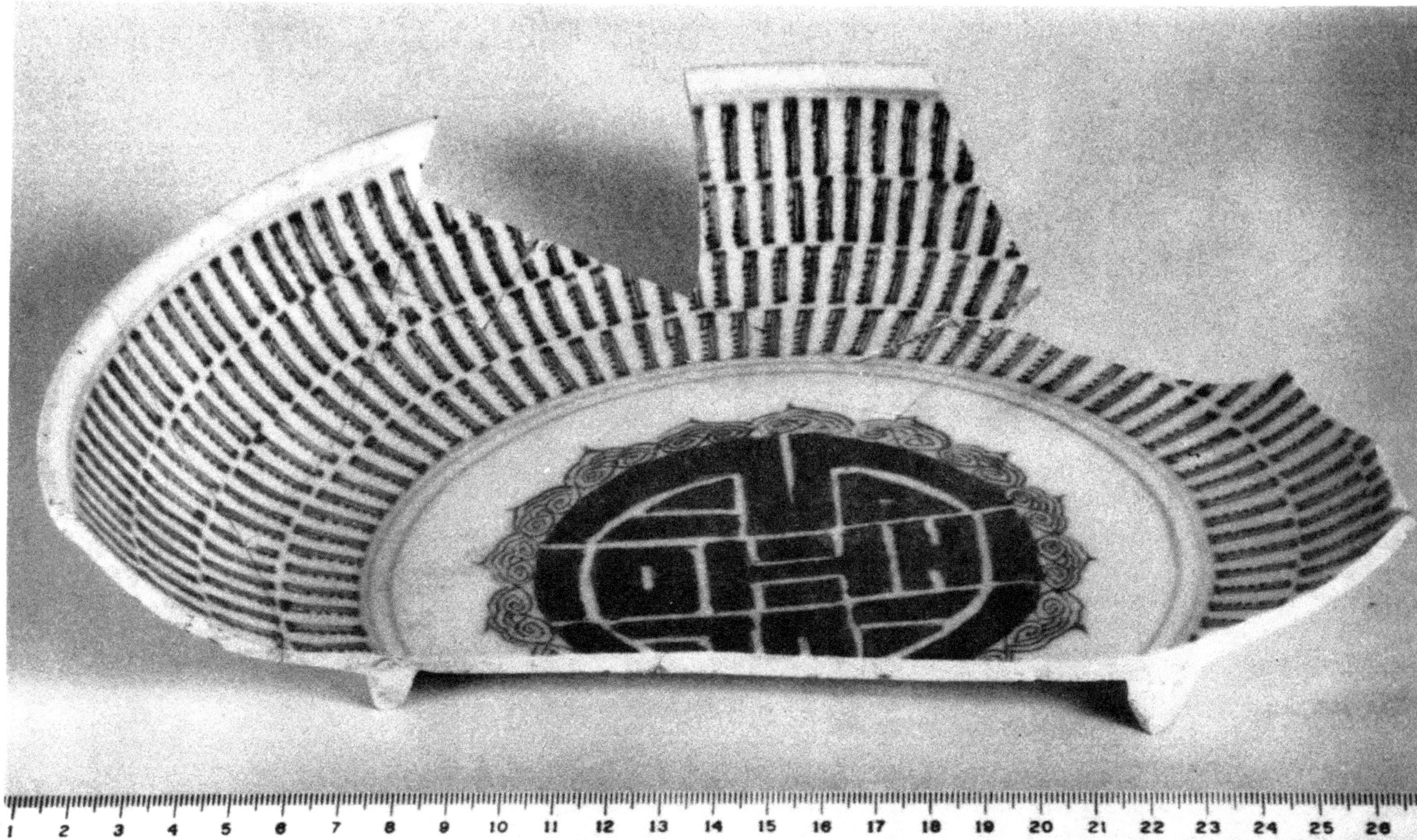

<u>Figure 12</u>: Blue-and-white ware, excavated deep in the Assay Office site. This ware pre-dates the start of the Chinese - American trade 1785, and was probably brought to America from Europe.

Figure 13: Section of a Colonial bateau excavated 10.5 metres below street level after the wharves were removed. (Assay site)

increasingly inferior quality. During this period, the Industrial Revolution in Europe led to the mass-production in factories of wares similar to Chinese export porcelain at a cheaper cost. Ships engaged in trade with China began to turn their attention to the European market, and by 1876, the China trade had virtually ceased. (McClure, 1962).

John Gardner, Curator of Small Craft, Mystic Seaport Museum, examined the remains of the two small craft and identified them according to the fastenings, joinery and methods used in hewing and working the timbers. Mr. Gardner believes that one vessel is definitely a Colonial bateau (Figure 13) and the other, possibly an English shallop.

The bateaux were an early type of American craft that were developed here and not in Europe. Bateaux were light-weight in construction. These boats were about 10 metres in length and designed to carry troops and could be dragged through shallow rocky streams or even portaged around rapids. Bateaux were continuously used for transportation and work, such as in log drives in Canada and Maine, well into the twentieth century. (Figure 14).

Bateaux were widely used in rapid flowing inland waterways. A fleet of bateaux was discovered by divers on the bottom of Lake George, New York, in the summer of 1960. (Gardner, 1966). Remains of the second vessel recovered was of heavier construction. The shape of the oak stem

Figure 14: Bateaux were an early type of American craft developed completely here. This light-weight craft was used extensively in inland waterways of the northeast. The above photograph shows a bateau being used in a log drive on a river in Maine during the early part of this century. (International Paper Company).

and frames and the three centimetre-thick pine planking suggest it was a shallop, a type of small craft more commonly used in coastal ports during the Colonial period.

We have only started to understand the story of the Old New York Port. There are many years of work ahead in recovering, analysing and conserving artifacts before we have an accurate picture of the lives, trades and crafts of residents, and visitors to the early sea port of New York.

REFERENCES

Broecker, W.S. and Kulp, J.S. 1957. "Lamont Natural Radiocarbon Measurements", Science 126:3287, 1324-1334.

Committee on Piers and Wharves, (CPW), Common Council Approved Papers. Municipal Archives, New York.

Dean, N., March/April 1985. "Manhattan's Mystery Merchant Ship". Wooden Boat, pp. 96-100.

Diderot, D., 1959. A Diderot Pictorial Encyclopaedia of Trades and Industry. Ed. by Charles C. Gillespie, Dover Publications Inc., Vol. I and II, Vol. II, 387-400.

Ewen, D., 1827-30. Waterfront Maps. New York. On file at Bureau of Topography, Municipal Building, Manhattan.

Foster, C. (no date). An Account of the Conflagration in the First Ward of the City of New York December 16, 1835.

Gardner, J., 1966. "Relics of Ghost Fleet are Small Craft Bonanza". National Fisherman, October.

Geismar, J.H., Dec. 1986. "Digging Into A Seaport's Past" Archaeology, pp. 30-35.

Greenhouse Consultants, Inc. 1983. Historical Documentary Report - Assay Office Site, New York, N.Y. Project #32-83-0003 September 1983. Prepared for HRO International Ltd., 126 East 56th Street, New York, N.Y. 10022.

Kearns, B. 1986. "Digging New York: What We Can Learn From Archaeology", Archaeology of New York City, Sixth Annual

Professional Archaeologists of New York City Symposium. Museum of the City of New York , May 3, 1986.

Loret, J. 1986. "Marine Archaeology in New York City". Summaries: "Cities on the Sea - Past and Present". First International Symposium on Harbours, Port Cities and Coastal Topography. Haifa, Israel, September 22-29, 1986.

McClure, J. 1962. Chinese Export Porcelain for the American Trade, 1785-1835. University of Delaware Press.

Minutes of the Common Council, MCC. 21 vols. New York 1917- 1930.

Peterson, A.E. and Edwards, G.W. 1917. New York as an Eighteenth Century Municipality. Parts I and II. Longmans Green Co., New York.

Ratzer, B. 1776. Plan of the City of New York. Jeffrey and Faden, London. (Depicts New York 1766-1767).

Riess, W.C. 1985. "The Ship Beneath the City". The Mariners' Museum Journal. Vol. XII: 3:2-7.

Shumway, F.M. 1975. Seaport City, New York in 1775. South Street Seaport Museum.

Solecki, R.S. 1974. "The 'Tijger', an Early Dutch 17th Century Ship, and an Abortive Salvage Attempt". Journal of Field Archaeology. Vol. I. 109-116.

Townsend, McCoun. The Island of Manhattan at the time of Discovery (1609) New York: Underpinning and Foundation Company, Inc. Engineers and Contractors 1909.

Ure, 1867. Ure's Dictionary of Arts, Manufactures and Mines, Ed. Robert Hunt, F.R.S. Longmans, Green, Co., London. Vol. I, II, III. Vol. II 836-859.

Williamson, W.M. Adriaen Block: 1611-1614; Museum of the City of New York, 1959.

WILLIAM M. MURRAY

The Ancient Harbour Mole at Leukas, Greece

INTRODUCTION

Colonized by the Corinthians in the second half of the 7th century B.C.E., Leukas became an important point on the coastal trading route between Greece, the shores of the Adriatic, Italy and ports further west. In this role, Leukas served for centuries as a regional centre of Corinthian influence in the Greek northwest. Because the city's importance was a function of its strategic position on a major trade route, the harbour at Leukas was an important element of the city. Despite this importance, every study of this harbour has been based on information gathered almost a century ago, long before snorkels, masks and fins were commercially available. Before a valid analysis of the harbour can be attempted, therefore, the current state of the ancient remains must be determined. To this end, two snorkel surveys were conducted in 1980 and 1983. They reveal a broad (10 metre-wide), submerged mole running some 600 metres from island to mainland at the south entrance to the Leukas straits. The placement and mass of its structure indicate its function as the main breakwater and loading dock for the southern side of Leukas' ancient harbour. Blocks still preserved in situ on the mole's surface suggest the sequence of construction and indicate that since its original construction, sea-level has changed relative to the mole +3.35 m (± 0.85 m). Pottery found along the sides of the mole, as well as historical and meteorological considerations, indicate that the structure was built before 428 B.C.E. and continued in use for many centuries. Even so, the latest pottery found here (dating to the 4th-6th centuries C.E.) may come from ships that ran aground on the mole after rising sea-levels put it out of use. If the South Mole at Leukas originally dates to the 7th century, it ranks as one of the earliest large harbour installations from ancient Greece and antedates the famous mole of Polycrates by more than 50 years. If excavation eventually proves that it dates to the 6th or 5th centuries, it will still remain one of the most impressive harbour moles of ancient Greece and a tribute to the skill and sophistication of the harbour engineers who built it. We are now in a position to reconstruct more accurately the true nature of Leukas' ancient harbour.

THE HARBOUR AT LEUKAS

A narrow, shallow, 6 km.-long stretch of water separates the island of Leukas from the northwest shore of the Plagia Peninsula on Akarnania's western coast (see fig. 1). These shallows extend from the northern sand spit past both the modern and ancient sites of Leukas and end at a submerged bank running from island to mainland (Fig. 2). This is

Figure 1: General map of Akarnania, western Greece.

referred to here as the South Mole; on many maps and charts, this submerged bank is called the "Corinthian Mole."

The history of the Leukas Straits is closely tied to its topography (Murray, 1982). When the shallows here were navigable, a safe shortcut existed inside Leukas for coastal traders, and offered a lucrative site for the growth of a city with a two-fold economic base -- agriculture and mercantile trade. According to the geographer Strabo (10.2.8), who lived during the time of Augustus, the Corinthians perceived the value of this site during the tyranny of Kypselos (657-625 B.C.E.) and settled it along with two other nearby sites (Anaktorion and Ambrakia). The compiler of the 4th century periplus (Ps. Skylax 34) adds that the Corinthians were called in by an earlier Akarnanian community to settle internal strife, but in the end, expelled the Akarnanians and held the site for themselves. These same Corinthians, according to Strabo, were the first to cut through an isthmus joining the island of Leukas to the mainland, and by inference, were also first to construct a harbour in the sheltered water east of their settlement. Even though Strabo does not mention the construction of a harbour at Leukas, it would be difficult to explain the considerable labor invested on opening the shipping canal if a harbour for the adjacent city was not also part of the plan.

Although the precise details of the pre-existing settlement, the nature of the isthmus, the canal cut through it, and the exact date of the Corinthian foundation come mostly from late sources and are impossible to verify, the general tradition of a Corinthian colony here is an undisputable historical fact (Oberhummer 1887). The site was well chosen, and the city prospered well enough to send 3 ships to fight the Persians at Salamis in 480 B.C.E. and in the following year 800 hoplites (including men from Anaktorion) to Plataia (Hdt. 8.45.1; 9.28.31). Compared with contingents from other Greek states, these contributions were not large, but nevertheless represent a degree of prosperity and commitment unmatched elsewhere along this coast except for the nearby Corinthian colony of Ambrakia (which sent seven ships; Hdt. 8.45.1).

Subsequent history shows that Leukas' prosperity continued to grow and that the city served for centuries as an important base for Corinthian influence in the Greek northwest (Rondoiannis 1980). Of all the site's advantages, the existence of a large protected harbour contributed most to its prosperity. The importance of Leukas lay mainly in its strategic position on the coastal route between the Corinthian and Ambracian Gulfs, and more than that, in its control of the Leukas straits. When this canal was navigable, a route existed inside the island of Leukas that was safer than the outer passage along the windward shores of the island's west coast (Murray 1982). Without its harbour and navigable canal to foster commercial development, it is likely that Leukas would have remained as it was before the Corinthians arrived -- a small insignificant coastal community.

It would seem, therefore, that as early as the archaic period, the harbour at Leukas was a critical component of an important city on Greece's western coast; and we have every reason to suspect that it

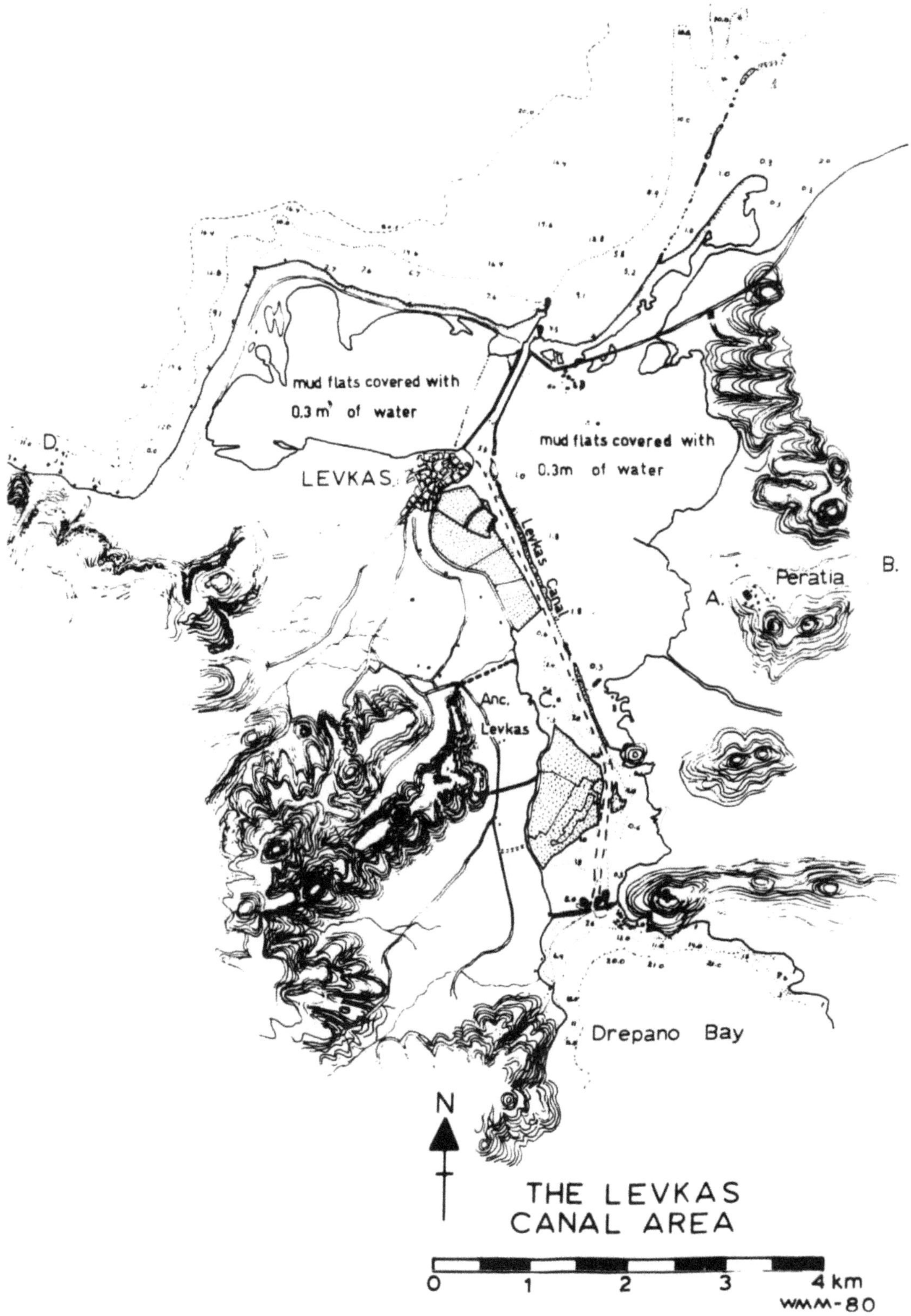

Figure 2: The Leukas Straits.

dates to the original Corinthian foundation of the 7th century B.C.E., or to some time soon thereafter. The South Mole, furthermore, (or a smaller one now covered by it) must have been a feature of this early harbour. There is a simple reason for this. From October to April, the frequency of S and SE winds is significant (Ginis 1974). Without some protective breakwater, any harbour in the straits south of the isthmus would be dangerously exposed during the winter months and potentially at risk for more than half the year. If one assumes that the Corinthians created a harbour when they cut the shipping canal, prevailing wind conditions dictate that a breakwater must have been constructed south of the city as part of their plan. In doing precisely this, we will see that they turned the southern third of the present-day straits into a protected anchorage -- a huge harbour by ancient Greek standards.

MODERN RESEARCH

The "official" rediscovery of this harbour occurred in the 19th century when the British Navy began systematically to survey this coast of Greece. The existence of a submerged bank running from island to mainland must have been known to any who used Port Dhrepano at the south end of the straits, yet it was not until the early 19th century that the mole first appeared on a chart (Smyth 1819?) A half-century later, the mole is mentioned by Captain Mansell who surveyed the straits in 1864 (Mansell 1864), and accordingly, is mentioned in the first edition of the Mediterranean Pilot (Great Britain 1880). A few years later the authoritative German scholar Eugen Oberhummer searched for this mole but wrote that he could not verify its existence (Oberhummer 1887). By 1904, however, the existence of a submerged mole at the south entrance to the Leukas straits was a generally accepted fact. This was largely due to the observations made when the modern canal was first dredged through the shallows here in 1902 (cf. the observations of Sakellaropoulos in Negris 1904; Partsch 1907). The opportunity to examine the lagoon silts to a depth of 5 metres sparked interest in the canal area and in succeeding years, scholars debated the implications of what was (or was not) found. The so-called "Corinthian Mole" at the south entrance to the modern canal occupied a central position in this debate.

A Greek geologist named Phokion Negris was among the first to explore the implications of the submerged mole, yet his observations were limited by having been made from a boat without the aid of any device to look under the water (Negris 1904). He therefore reported the depth of the mole as a general 2.5 m but went no further than to suggest a rise in sea-level of at least 3.5 m since its original date of construction (which he placed around 600 B.C.E.) One year later, a German geologist named Walter von Marees carried out a full geological survey of the canal area for Wilhelm Dörpfeld (Marees 1907). He found that the mole did not possess an even surface and reasoned that its blocks had been knocked apart by the waves; he concluded, therefore, that the mole provided ambiguous evidence for any precise change in sea-level.

Marees added a few more details to our picture of the mole (e.g., that its entrance width was 80 metres -- it is, in fact more like 60), but he was not really interested in a comprehensive study of it. The main interest it held for scholars of this period was as an indicator of ancient sea-level. These were the days of the so-called "Ithaka-Frage," that is, whether modern "Ithaki" or "Lefkada" should be identified with Homer's Ithaca, and the mole was examined mainly to provide information concerning the nature of Leukas as an island in antiquity. Thus, no in-depth account of the mole ever appeared and no concrete evidence was ever produced for dating the structure other than its obvious association with the nearby site of ancient Leukas. Indeed, Lehmann-Hartleben, whose discussion of the harbour is the most extensive one in print, concluded from the same evidence (provided by Partsch, Negris and v. Marees) that the South Mole was constructed primarily as a silt barrier in the third century B.C.E. (Lehmann-Hartleben 1923). What, therefore, is one to believe concerning this important harbour on the coastal route between Greece and Italy? Because the true nature of the ancient remains had never been adequately determined, meaningful conclusions were impossible. This gave rise to simple speculation concerning the harbour's period of use, its appearance and the original reasons for the construction of the South Mole.

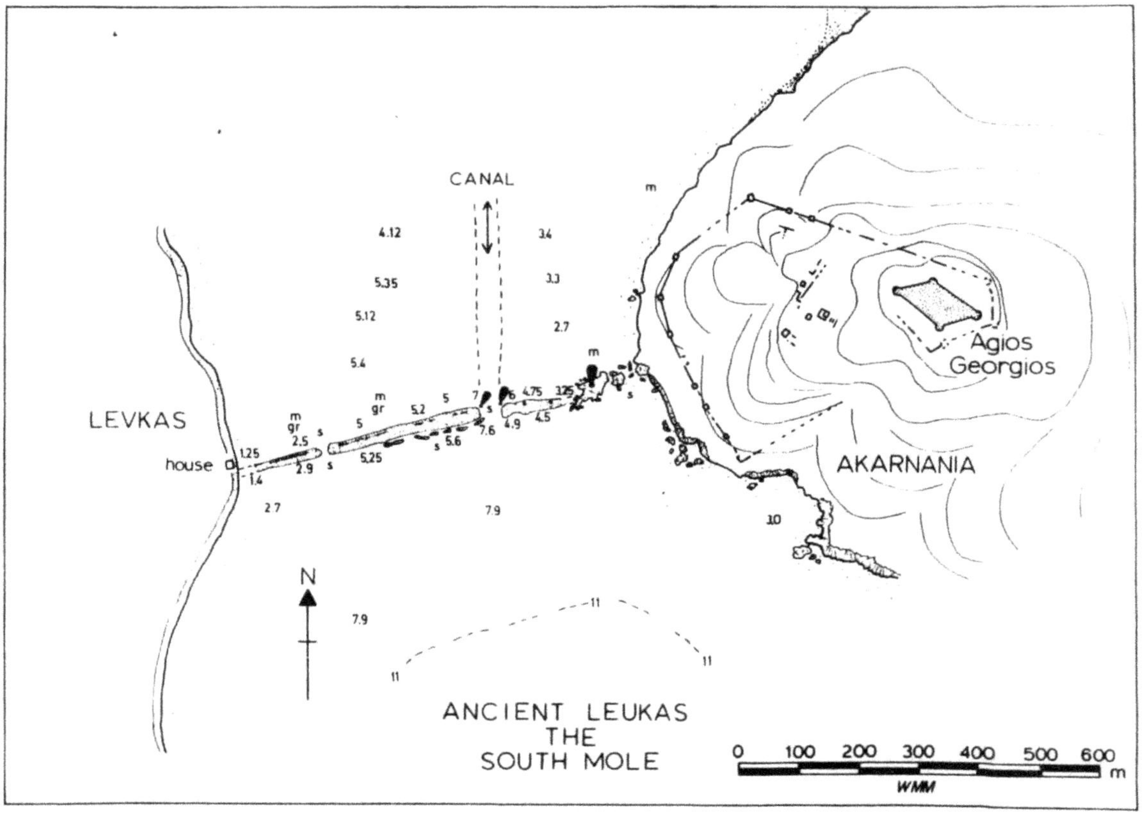

Figure 3: Southern end of the Leukas Straits.

THE SURVEYS OF 1980 AND 1983

In the summer of 1980 a survey of the South Mole was conducted as part of a wider study of western Akarnania (Murray 1982). The principal aim was to determine the extent of the ancient remains. I found them to be considerable, stretching from the small islet called Volios near the Akarnanian shore to a small pink house, opposite, on Leukas (Pl.1, Fig. 3). The ancient mole is composed of two sections, the eastern one being the shorter (138 metres from its western end to the islet), the western one the longer (401 metres from its eastern end to the shore), with a gap of approximately 60 metres in between. Though the original surface is largely gone, numerous blocks still remain in situ. These were first thought to be paving slabs from the mole's original surface (Plate 2). A revised opinion of these blocks is discussed at length below. After some 17 hours of investigation, it appeared that too much remained over too wide an area to be recorded accurately without help. Therefore, a methodology was developed for an intensive snorkel survey and the work was completed by a team of divers in 1983. The observations following result from about 250 man-hours in the water.

The team which conducted this survey was composed of Suzanne Murray, John Maseman, Erich Priester, Adam Cohen, and Charles Hedrick; Buddy and Marilyn Baker helped for a few days on the western arm. The following report is the direct result of their aid, and endurance. More work remains to be done on the South Mole, but it would require SCUBA equipment, dive boats and a considerable support team.

THE SOUTH MOLE

The mole is oriented along a more or less straight line running some 600 metres E to W. Measuring such a structure is theoretically simple, but as with all in-water surveys, practical measurement is more difficult to achieve. In order to avoid wasting a lot of time while in the water, a strict procedure was worked out on land and then applied first on the simpler eastern arm. By the time the western arm was reached, our routine was well established. Efficiency was necessary here because our base camp was located hundreds of metres from the mole on the Akarnanian shore and the commute time increased as we proceeded westward.

Our method involved first laying a continuous baseline along the surface of each arm. For this purpose, a bright yellow, light-weight line was laid along the top of the mole and weighted down with rocks to diminish the inevitable dislocations caused by fishermen's hooks. Buoys placed at points along these baselines allowed their positions to be recorded with a transit from the shore and plotted on a 1:4000 survey made by the British Navy in 1904 (Aylmer 1904). This provided us with the backbone upon which all additional measurements could be placed.

Our next task -- the measurement of each arm's length -- was made quite easy by the existence of the continuous baseline. Starting from the eastern shore, we laid down a 50 metre line marked in metre intervals. Unfortunately the line used floated, so that it had to be weighted with stones. To ensure that the line did not stretch, we checked it repeatedly with a 50 metre tape (made of fibreglass). Every 5 metres, long nails were driven into the mole's surface to which bright yellow tags had been affixed, clearly marked with the distance from shore (5,

Plate 1: Area of South Mole from the East

Plate 2: In situ blocks of South Mole, western arm.

10, 15, 20 etc.). These immovable points allowed us to orient ourselves quickly wherever we were on the mole and were necessary during periods of low visibility. Ironically, a modern harbour mole was under construction at nearby Lygia on Leukas, and when the tidal currents flowed northward, the area of the South Mole was inundated with silt. On a few days, we were forced to stop working because of low visibility. Fortunately for us, an ancient amphora was recovered near the site of the new mole; work there was temporarily suspended and the water cleared up markedly.

When the visibility allowed, a fixed number of width and depth measurements was taken at each of the 5 metre intervals designated by a distance marker. Our aim was to define the overall shape of the mole, to record a rough depth profile of the submerged structure and to measure the relative positions and depths of all worked or in situ blocks preserved on its surface.

Furthermore, most of our depth measurements were taken early in the day when the sea surface was flattest. Once these measurements were completed on a segment, the 50 metre line was repositioned along the baseline toward the west and the process of measurement was repeated. Since we also wanted to date the mole's period of use, we kept an eye out for fragments of discarded transport amphoras. Broken during the process of transfer from ship to shore, and kicked into the sea with a curse, these discarded fragments provided priceless information on the life-span of the harbour mole. Diagnostic fragments were noted on the survey before removal from the water and after being drawn and photographed were returned to the places where they were found. The mass of information resulting from this routine enabled a fairly clear picture of the mole to emerge. The information related to its original appearance, the methods by which it was constructed (and later dismantled) and its long period of use.

The Western Arm. Clearly, the most important information comes from the western arm of the mole where long lines of worked blocks remain seemingly in situ on the north (Pl. 2) and a "distinct face" (see below) is discernable along the south side of the mole's surface. When the positions of the individual in situ blocks and sections of distinct face are plotted in relation to the baseline, it is clear that both features run more or less in two straight lines 10 metres apart and form the approximate north and south edges of the mole's original surface (Fig. 4). It is clear that on its north side, the mole presented a straight face to the water. On the south, the evidence is less clear and requires some interpretation.

Here, a curious scatter of rubble and worked blocks between the 200 metres mark and the end of the western arm provides telling evidence. Too deep to be a wave-catcher or prokymia of the type seen at Caesarea Maritima (Oleson 1984), this spill of rubble looks more like the refuse from those who quarried away the surface of the mole in late antiquity. The presence of numerous worked blocks in the spill indicates that this side of the mole also originally exhibited a perpendicular face, either descending into the water as on the north, or set back from the water as some sort of sea wall. At some time, however, this side was systematically quarried for its rectangular blocks, which were all wrenched from their original positions. This would explain two features of the south side: (1) the distinct drop at the edge of the mole's surface -- what I call a well-defined or distinct face, and (2) the current haphazard positions of numerous worked blocks along the south flank of the mole and in rock spills to the south, separate from the mole's main mass. Similar rock spills are not apparent on the eastern arm because the area is more silted-in and the region adjacent to the mole is covered with thick grass. But worked blocks scattered haphazardly along its margins hint that the eastern arm originally looked like its partner to the west. The following reconstruction of events is likely:

<u>The Sequence of Destruction and Construction.</u> At some point after rising sea-levels put the mole out of use, masons with hammers and chisels removed the central, paved surface of the platform on both arms of the mole. Evidence for this surface can still be seen in a few large paving slabs dumped well to the north of the worked-block line at the 298 metre mark on the western arm; another slab is located at the 237 metre mark. Relatively speaking, these slabs are huge, measuring roughly 2.0 x 1.6 metres, and it is hard to imagine what purpose they might have served other than as paving slabs. By chiselling them free, the masons were then able to get at the facing blocks along both margins. The numerous blocks now scattered along the south side of the mole imply that they were removed in this direction, i.e., toward the south, at least on the western arm. Boats or barges anchored along this side might then have winched them into the water for transport to shore. Easier to move once in the water, but less easy to control, some blocks may have slipped from their ropes and now lie where they fell to the bottom along with the rubble chipped from the mole's interior. At a later time, sand filled the cavity between sections of the spill and the mole's south edge and now gives the impression, in some places, of a structure separate and distinct from the mole's south face. If I am correct, the cavity exists because the boats' tackle swung the blocks away from the south face. More than likely, if the sand were dredged away, a continuous rock spill would be visible, thinning as it continued to the south.

As for the mole's north face, numerous blocks are preserved on the western arm in their original positions (cf. Table 1 and Fig. 4); two courses are visible in the 148-239 metre section and it is even possible that a third can be unravelled from the measurements recorded at the 51-59 metre section. Here, the depth is on average 1.4 m, as opposed to 2.1 m for the upper course at the 148-239 metre section. Are these blocks from a third course, or was the original surface of the mole simply higher out of the sea near the shore? Excavation of the mole's face here might allow us to say for certain. It is curious, however, that these blocks near the shore and at a higher level were not robbed.

One final observation should be presented before turning to the mole's sequence of construction. A significant mound of rubble is deposited north of the worked block line on sections of the western arm. If this

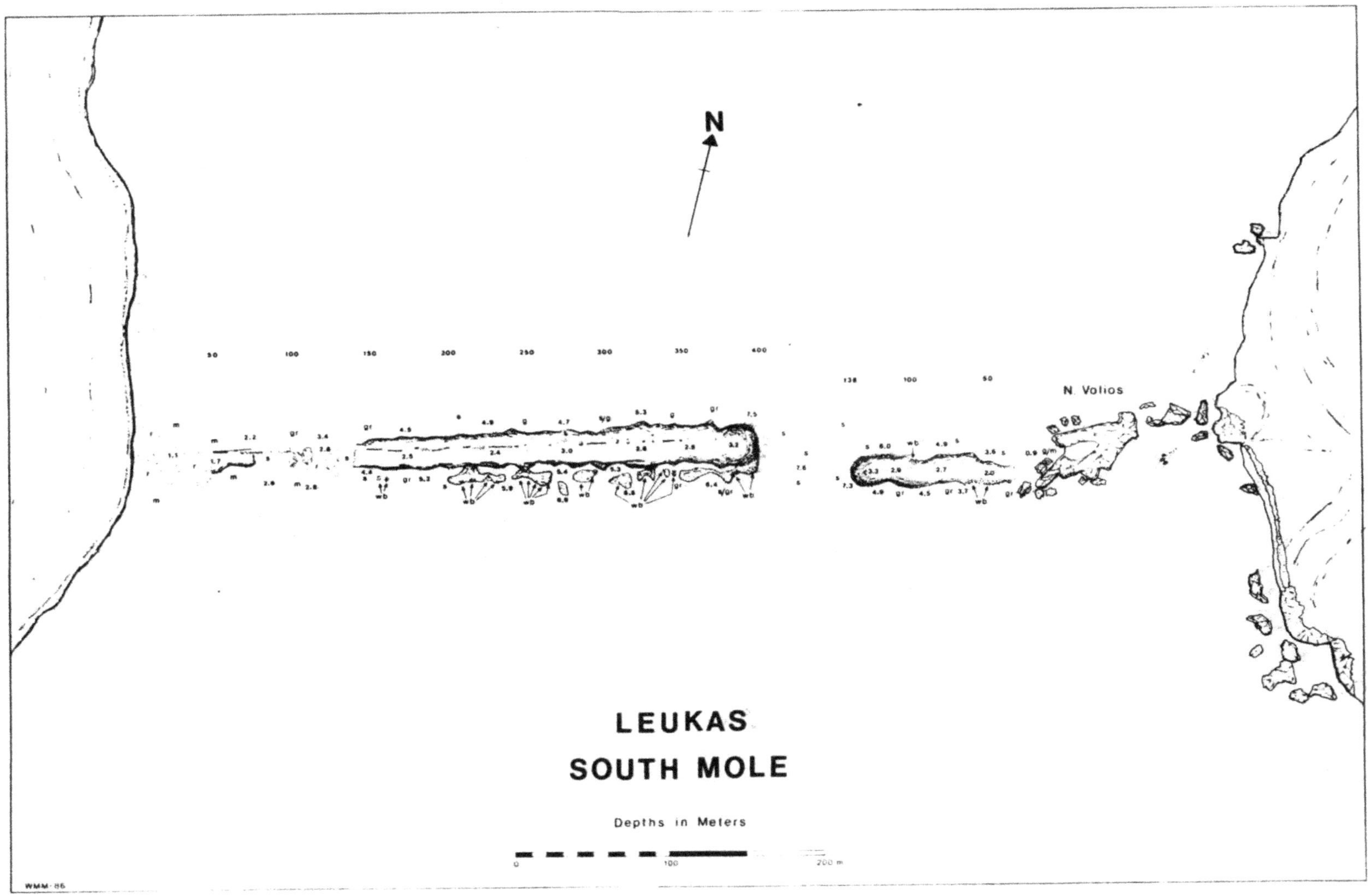

Figure 4: The South Mole at Leukas.

line is correctly interpreted as the original northern edge of the mole's surface, the considerable mound of rubble currently located to the north of these blocks must be explained. That the general depths to the north should be less than those to the south is not surprising. The mole hinders the southward flow of silt, washed from the nearby mountainsides, which has accumulated over time along the mole's north side, but this does not explain the mass of rubble which continues far north of the worked block line. Perhaps those who robbed the south face were responsible for this debris. If barges aiding in the mole's destruction were anchored to the south (as is implied by the rock spill), one might expect most of the rubble taken from the surface and interior of the mole to be dumped off its north margin.

Working backward from the destruction process, two distinct steps in construction can be noted. First, a substructure of rubble composed of rock and earth was dumped into the water at the southern entrance to the straits -- a simple technique, still in use today throughout Greece. A 60 metre gap was left between the arms of the mole to allow passage into and out of the harbour. Once a flat platform some 10 metres wide was achieved at the surface, the north and south margins of the mole were cut back or notched just below sea-level to receive large rectangular blocks, laid end-on to the force of the waves in at least two courses. This technique, well known in the eastern Mediterranean, added strength to the platform and gave the dock a face that descended straight into the water allowing ships of appropriate draught to tie up close alongside. The interior surface of the mole was built up to the upper level of the blocks along the margins and then paved with large slabs.

Evidence for Change in Sea Level. The following table presents the depths recorded from *in situ* worked blocks located on the western arm. An analysis of these data follows Table 1. The depths are listed in sequence from west to east. Thus, the first value given in a sequence of numbers is the western-most of a series of depth measurements. Unless otherwise noted, all worked blocks are located to the north of the baseline.

TABLE 1. Worked blocks locations and depths.

Abbreviations used in Table 1:

D from L	=	distance in metres along baseline from Leukas
D to BL	=	distance in metres from worked blocks to baseline
wb	=	worked block
wbs	=	worked blocks
i.s.	=	*in situ*
n.i.s.	=	not *in situ*
a of t	=	measurement taken from the centre point of the block's axis of tilt
(S)	=	block located to the south of the baseline
2c	=	second course
1c	=	first course

D from L	D to BL	Depths
51-60	at BL	1.4, 1.3, 1.37, 1.37, 1.37, 1.35 1.4, 1.4, 1.37, 1.4, 1.45, 1.4; blocks i.s.
65-66	1.7	1.6, 1.67, (a of t); blocks i.s.
68		1.7 plus two more wbs n.i.s.
68-88		a loose line of wbs, greatly disturbed
88	at BL	1.95; i.s.
93-101		disturbed line of wbs
110	at BL	2.35
111	at BL	2.35
148-51	1.0	2c: 1.78*, 1.7 (S), 1.75@, 1.83, 1.73, 1.7 1c: 2.4*, 2.32@; i.s., but slightly disturbed

* and @ indicate blocks set on top one another

D from L	D to BL	Depths
151	1.0	1c: 2.45 i.s.
153-58	1.0	1c: 2.55, 2.65, 2.7, 2.52, 2.5; blocks i.s.
155-58	1.0	2c: 1.82, 1.85 (a of t), 1.75 (a of t), 1.93 (a of t), 2.0; blocks i.s.
160	1.0	1.87 i.s.? wbs on both sides; wbs to south, n.i.s
162-63	at BL	1c: 2.3, 2.25, 2.4; blocks i.s
163-67	at BL	2c: 1.85, 1.85, 1.77, 1.77; blocks i.s.
167	at BL	1c: 2.13; i.s.
169-71	at BL	2c: 2.3, 2.12, 2.3; blocks i.s.
171	at BL	1c: 2.23, i.s.?
180-82	c 1.5	2.7, 2.67, 2.7, 2.65 (a of t); blocks i.s.
182-91		disturbed surface
187		2.8; i.s.?
188		2.6; i.s.?
191-99	at BL	2c: 2.17, 2.2, 2.17, 2.2, 2.2, 2.12, 2.15, 2.25, 1.9; blocks i.s.
201	1.9	1c: 2.55, 2.55, 2.5, 2.55; blocks i.s.
201-209		disturbed line of wbs; general depth between 2.9 and 3.0
210	2.0	1c: 2.75; i.s.
210-14	2.2-2.3	2c: 2.3* (a of t), 2.5 (a of t), 2.57, 2.57, 2.5 (a of t), 2.6

* this block definitely from the second course, i.s.

D from L	D to BL	Depths
214-26		jumble of wbs located north of the BL
227-30	2.1-2.3	2c: 2.45 (a of t), 2.35, 2.37, 2.35; blocks i.s.
230-33	2.3	1c: 2.9, 2.95, 2.8; blocks i.s.
233-39	2.3-2.1	2c? 2.1*, 2.3, 2.4 (a of t), 2.4, 2.3 (a of t) 2.15, 2.15, 2.2; blocks i.s.

* this block displaced on its side, but otherwise i.s.

D from L	D to BL	Depths
241-42	2.4	2.7, 2.55; blocks n.i.s.?

243-45	3.0	six wbs, n.i.s.
250	4.5	a few wbs, n.i.s.
251	just south of BL	2.35; i.s.?
265-67	3.8	2.45 (nis), 2.7 (a of t), 2.65 (a of t), 2.5 (a of t); three wbs in a row, i.s.?
271	4.71	2.8; one wb under a jumble of disturbed blocks (i.s.?)
275	5.0	2.8 (i.s.?); a jumble of wbs surround it
276	c 10.0	a small, round, flat worked stone
282-87	2.8	a jumble of wbs
287-90	2.8	2.5, 2.35, 2.5, 2.45; four wbs in a line (i.s.?)
290	3.6	one wb, n.i.s.
290-302	4.0	2.75, 2.8, 2.73, 2.65, 2.65, 2.6 (a of t), 2.65 2.5, 2.65, 2.7, 2.65; blocks slightly tilted, but i.s.
306	7.7	2.95; one wb, i.s.?
309	4.1	2.55, 2.7, 2.7; other wbs jumbled nearby
314		two large wbs, n.i.s.?
316	4.2	2.45 (a of t), 2.45; i.s.?
	5.8	2.6 (a of t); i.s.?
327	4.2	2.6, 2.5, 2.4 (n.i.s.); two wbs i.s.? jumbled wbs to the north
332	4.5	2.5, 2.6, 2.5; i.s.?
	8.0	2.55 (a of t), 2.2; n.i.s.?
338-46	4.8-5.1	2.45, 2.4, 2.55, 2.6, 2.55, 2.45; more than six wbs; those recorded here are i.s.
339	8.5	2.5; one large wb, n.i.s.
349-51	4.9	2.6, 2.7, 2.7; three wbs, i.s., but askew
355	6.0	2.8; one wb, i.s.?
373	5.9	2.75; four wbs, i.s.?
374	c. 4.0 m (S)	one wb, n.i.s.?

Analysis of Table 1. The mass of depths presented in the above table can be separated into three different levels. Because <u>in situ</u> blocks remain in two superimposed courses at the 148 metre mark, we can be fairly certain that the varying depths correspond to different courses. Generally speaking, the depth of the lowest (or first) course ranges between 2.3 and 3.0 m, while that of the second course is between 1.7 and 2.5 m. In the section where the two courses overlie one another (148-51 metres), the depths are approximately 2.4 and 1.8 m. Elsewhere, where two different levels of blocks can be discerned in the same area, the average depth of the lower course is about 2.5 m, while that of the upper one is approximately 2.1 m. The last value can be interpreted as the approximate level of the mole's original surface, unless there was a third course as indicated by the 1.4 m depths close to the Leukas shore (in the 51-60 metre section).

It is unlikely that the surface of such a dock intended to service medium to small-sized "coasters" stood at less than 1.0 m, or more than 2.0 m above the water. Using the values discussed above for the original surface of the mole, the maximum and minimum change in sea level can be computed as 2.5 m and 4.2 m respectively. The minimum value stems from the depth of the topmost course -- 1.4 m -- plus 1.0 m, plus 0.1 m (for variations in depth values caused by tidal changes). The maximum value stems from the depth of the second course -- 2.1 m -- plus 2.0 m, plus 0.1 m for tide error. The average of these two values gives us the change in sea-level indicated by this mole as +3.35 m (± 0.85 m).

The Date of Construction and Period of Use. The end result of the construction process was a solid structure that served the harbour area a long time. If we judge from pottery found alongside the mole (to be discussed in a separate article), the harbour at Leukas was in use for about a millennium, from the 5th-4th centuries B.C.E. to the 4th-6th centuries C.E. (Murray 1982). This evidence, however, must be used with caution for it applies primarily to the harbour and not necessarily to the South Mole. Since the mole is clearly a man-made structure, no one can doubt that pottery found on the mole's surface provides us with a terminus ante quem for its date of construction and consequent use. And since the earliest evidence would have been covered over by later deposits, we might also expect the mole to antedate the earliest surface pottery currently visible. Only excavation of the deposits along the mole's surface and sides, therefore, can provide conclusive evidence for the mole's original date of construction. We can assert on this evidence, however, that the mole was built and in use at least as early as the fourth century B.C.E.

Historical considerations lead to the conclusion that the South Mole was constructed even earlier than the surface pottery indicates. An examination of the hill adjacent to the mole's eastern end (cf. Fig. 3) reveals the poorly preserved remains of a fortified settlement (Murray 1982). Today few of the blocks from the settlement's fortification wall remain in situ, but its plan was fortunately recorded by Dörpfeld who carried out excavations here in the early 1900's (Dörpfeld 1927). Considering the position of this town adjacent to the South Mole, it was probably fortified by the Leukadians to protect the mainland side of the mole from the Akarnanian descendants of those driven out when Leukas was colonized in the 7th century. There can be no other reason for this town's existence. Indeed, such an interpretation fits perfectly what we know of a town called Nerikos, first mentioned by Thucydides in 428 B.C.E. (Thuc. 3.7.4; Murray 1982). If correct, the placement of Nerikos proves the existence of the South Mole during the Peloponnesian War. It would also support what the prevailing wind conditions and the early prosperity of Leukas have already suggested concerning the harbour here, i.e., that the South Mole was probably built at the same time as (or soon after) the shipping canal was cut through the isthmus. On the evidence of this town, therefore, it appears that the mole was constructed by the last third of the 5th century at the very latest.

As for the date of the mole's abandonment, pottery evidence is less informative. The harbour adjacent to the mole was clearly in use for more than a millennium, but the mole is a different matter. It would seem inconceivable that this structure could have remained in service for nine or ten centuries without needing major repairs. And yet, our survey turned up no clear evidence for such repairs. One might have suspected that the third course at the 50-59 metre section represents some sort of repair, or perhaps even the raising of the mole's surface to offset rising sea-levels. Yet this course is composed of blocks made of similar stone and cut to similar sizes as those found elsewhere in the first two courses. Without more evidence, it must be concluded that if there were repairs, either they were not extensive, were robbed out themselves, or were executed in a manner similar to the original method of construction. Stylistically, what remains of the mole today appears to be from one construction phase. It is this fact, above all else, that leads to the conclusion that the mole must have been abandoned by the date of the latest pottery.

When rising sea-levels put the mole out of use as a dock, it increasingly became a hazard to navigation. As eventually happened with Herod's great mole at Caesarea Maritima, the South Mole began to act more as a reef and less as a breakwater. Some of the latest pottery must have come from unfortunate ships driven onto the mole and holed. This would help to explain the incredible span of time indicated by the pottery found along the sides of the South Mole. It is thus impossible to determine from pottery evidence alone the precise date when the mole itself was finally abandoned. We can be fairly confident, however, that by the date of the latest pottery found here (4th-6th centuries C.E.), the surface of the South Mole was close to the water or just awash.

CONCLUSION

With a total length of some 600 metres, and a surface width of 10 metres, the South Mole at Leukas ranks among the most extensive harbour installations of ancient Greece. It compares favourably with the archaic moles at Delos (280 m long, 4-5 m wide, in 2-3 m of water), Eretria (600-700 m long with 20 m of water at its end), Eleusis on Thera (south mole 360 m long; north mole 110 m long). Only the south mole at Samos (370 m long, but reportedly in 40-60 fathoms of water at its end = 73-110 m!) and the mole at Histiaia (c 900 m long) are demonstrably more massive (Lehmann-Hartleben 1923). If dated originally to the period of Corinthian colonization in the 7th century, the South Mole at Leukas would rank, therefore, as one of the earliest and largest examples of harbour works in Greece, antedating the famous mole of Polycrates at Samos (cf. Hdt. 3.60.3) by more than half a century. Even if future excavation reveals the mole to be younger, the South Mole at Leukas will remain an impressive example of Greek harbour engineering which withstood the sea for centuries before rising sea-levels put it out of use. The results of the 1983 survey provide new information concerning this harbour which will enable a more accurate assessment of its role in the history of an important Corinthian colony.

REFERENCES

Aylmer, Lieut. S.H.J. 1904. Canal Santa Maura Road to Port Drepano. Survey by HMS "LEANDER" B 9290 Shelf 16r. Archives of the British Admiralty, Taunton, Somerset.

Dörpfeld, W. 1927. Alt Ithaka. 2 Vols. Munich.

Ginis, S. Ch. 1974. Ai Anemologikai Synthekai tou Ioniou Pelagous. Diss. Athens.

Great Britain 1880. Mediterranean Pilot. 1st edition. Vol. III. London.

Lehmann-Hartleben, K. 1923. Die antiken Hafenanlagen des Mittelmeeres. Klio Beiheft 14 (Neue Folge, Heft 1). Leipzig.

Mansell, Capt. 1864. Santa Maura, Plan of the Roadstead and Port Alexandro. D 8032 Shelf Sw. Archives of the British Admiralty, Taunton, Somerset.

Marees, Walther von. 1907. Karten von Leukas. Beiträge zur Frage Leukas-Ithaka. Berlin.

Murray, W.M. 1982. The Coastal Sites of Western Akarnania: A Topographical-Historical Survey. Diss. Univ. Microfilms International.

Negris, Ph. 1904. Vestiges antiques submerges. Ath. Mitt. 29: 354-60.

Oberhummer, E. 1887. Akarnanien, Ambrakia, Amphilochien, Leukas im Altertum. Munich.

Oleson, J.P., R.L. Hohlfelder, A. Raban and R.L. Vann. 1984. The Caesarea Ancient Harbor Excavation Project (C.A.H.E.P.): Preliminary Report on the 1980-83 Seasons. Journal of Field Archaeology 11: 281-305.

Partsch, J. 1907. Das Alter der Inselnatur von Leukas. Petermanns Geographische Mitteilungen.

Rondoiannis, P.G. 1980. Istoria tes Nesou Leukados. Vol. I. Etaireia Leukadikon Meleton. Athens.

Smyth, Capt. W.H. 1819(?) Plan of the North Anchorages of Santa Maura. Plan adapted from an unknown original. H. 366 Med. Folio 2 and L 1498 Med.Folio 2. Archives of the British Admiralty, Taunton, Somerset.

TINA M. NIEMI, EBERHARD A.W. FINKE

The Application of Remote Sensing to Coastline Reconstruction in the Argive Plain, Greece

ABSTRACT

Landsat-1 coverage of the Argive plain, Greece, is examined to address whether units observed on the satellite image can provide specific information about the paleogeography of the coastal area. Boundaries identified on the Landsat image and the computer-generated land cover map were compared to the Holocene stratigraphy known from subsurface investigations. From analysis of the Landsat-1 spectral data and generation of a computer-classified land cover map, areas with similar spectral reflectance properties were identified and mapped. There is a sharp boundary between the agricultural-rich alluvial valley and the coastal plain. A change in vegetation across this boundary enables us to detect it on the satellite image.

The Holocene stratigraphy is characterized by a lower late Pleistocene-early Holocene sequence of soils overlain by marine to lagoonal silts and clays and Neolithic to recent alluvium and soils. The inland limit of the marine to lagoonal muds marks the maximum paleoshoreline of a transgression. Landward of a beach barrier which developed during the transgression, peat dating to the mid-Holocene was deposited in a lagoon. The shoreline has subsequently prograded seaward by the infilling of the lagoon and deposition of alluvial sediments. The floodplain/coastal plain boundary seen on the satellite image correlates to the inland limit of the subsurface subaqueous mud unit, marking the maximum early to middle Holocene transgressive shoreline.

INTRODUCTION

> "The plain of Argos was apparently in early prehistoric times a bay running far inland; this was gradually filled up by the deposit of the numerous streams descending from the surrounding hills."
>
> Heinrich Schliemann, Tiryns (1885, 11)

The ruins of ancient coastal settlements now stranded from the seashore have provoked a long tradition of study of the Holocene coastline shifts in the Mediterranean. Rapid sediment accumulation on the coastal plains of Greece causes coastal progradation, thus preserving marine deposits under young alluvial fill. Recognition and investigation of these buried marine deposits allow for geologic reconstruction of the environmental setting of coastal archaeological sites.

Coring is a common and successful technique to determine the stratigraphy of Holocene coastal sediments. Some examples from Greece utilizing this method are discussed by Baetemann (1985), Finke (1984), Gifford (1985), Kraft (1972), and Kraft and others (1977). When cores are aligned in a traverse, a two dimensional section of the coastal deposits can be drawn. From the cross section, the relationship and character of the sedimentary units are described and used to interpret the environmental history. For example, the extent and thickness of buried marine deposits indicate former transgressions, soil horizons reflect land stability, and alluvial deposits record times of enhanced land erosion. Samples from the cores can also be analyzed to determine the past physical and biological environment, including fauna, flora, water temperature, current activity, and salinity.

In this paper, however, we introduce an approach combining both remote sensing analysis and subsurface core data to reconstruct past coastal environments of the Argive plain. From photogeologic interpretation of a Landsat image, several geomorphic and geographic features are defined. Then analysis of the spectral data allows us to produce a computer-generated land cover map which emphasizes areas with varying surficial cover. Finally, we compared the map produced from the satellite data with the Holocene stratigraphy known from subsurface investigations to test the results of the remote sensing techniques.

The Argive plain in the Peloponnese region of southern Greece is an especially suitable location for an investigation of the application of remote sensing (satellite) to shoreline reconstructions because of extensive historical data, archaeological and geological work. Data presented here are part of an ongoing investigation of the environmental history of the Argive plain.

STUDY AREA

The Argive plain is a broad alluvial valley with an areal extent of 243 square kilometres (Figure 1). The valley is seated in a graben between two geologic units. To the west, the Artemision-Ktenia Mountains are composed of Mesozoic limestones and cherts of the Pindos zone. To the east, Cretaceous flysch and limestones make up the Arachnaion mountains of the Pelagonian platform.

Two streams traverse the west side of the valley. The perennial Erasinos River is fed by springs in the limestone caves at Kefalari.

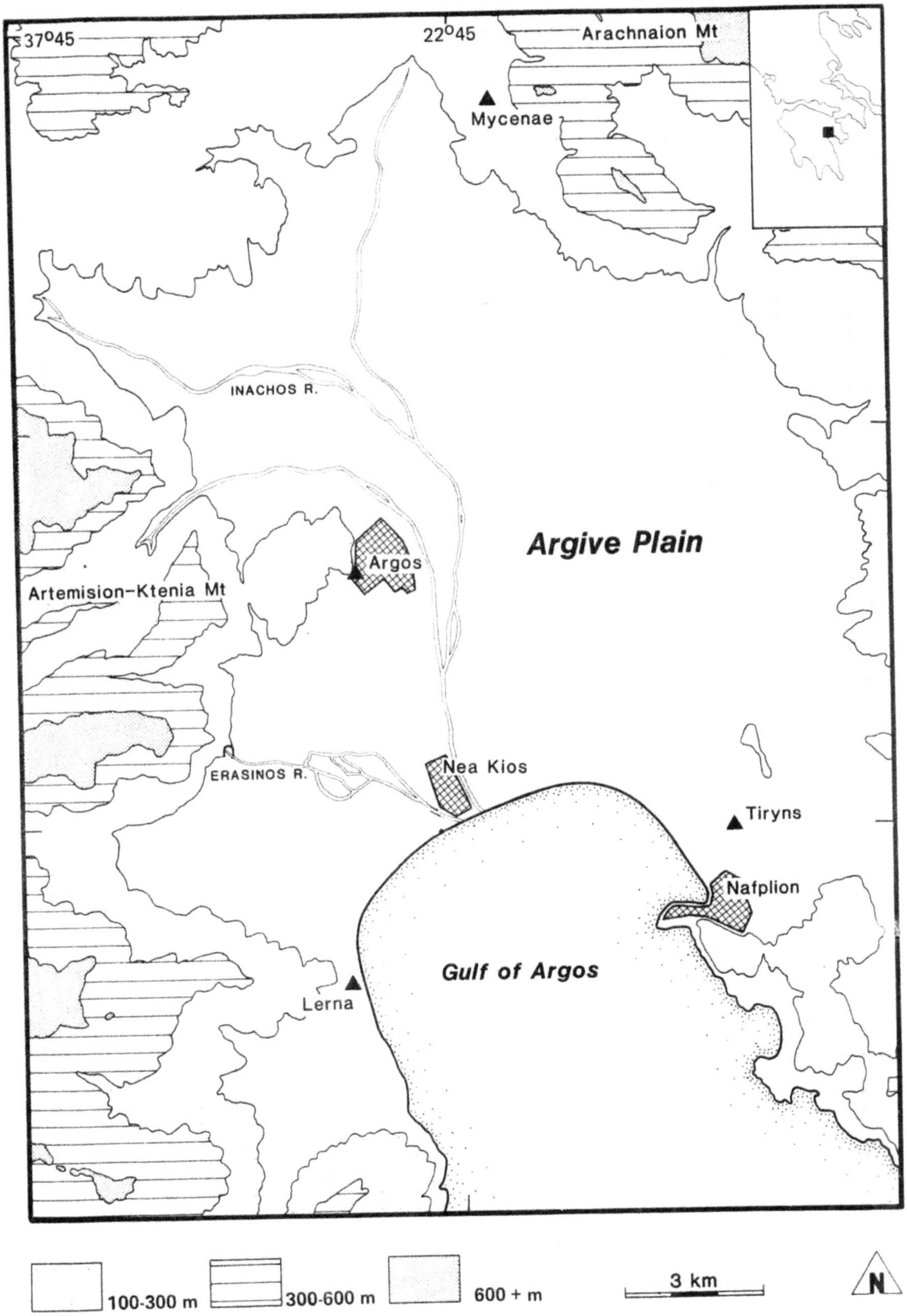

Figure 1: Location Map of the Argive plain in the Peloponnese region of southern Greece illustrating the major geographic features in the study area. Triangles represent major achaeological sites and cross hatched pattern marks the location of modern cities.

The sediment load of the Erasinos River is negligible due to its origin from water draining through fractures and caves in the limestone mountains. A seasonal stream, the Inachos River, carries most of its sediment load during flood stages. Major floods occur about once every 15 years. Sediments are either deposited on the valley floor, or are carried to the Gulf of Argos.

A broad arcuate coastal zone with very low relief separates the floodplain from the Gulf of Argos. Deposits of the coastal plain have a very high preservation potential because of the rapid sediment accumulation. Beach sand and gravel near the mouth of the Inachos River protect a portion of the coastal plain from wave action. At present, the beach barrier effectively ponds water in a small swamp near Lerna.

HISTORY AND PREVIOUS WORK

The coastal region of the Argive plain has a rich cultural history. The Neolithic (6000-3000 B.C.E.) and Bronze age site of Lerna rests upon a gentle cliff, west of the Gulf of Argos. The site of Tiryns, a Bronze age (3000-1200 B.C.E.) sister city of Mycenae, is located on a limestone knoll, 1.5 kilometres east of the present shoreline. These ancient ruins and others in the Argive plain, have given rise to century-old speculation regarding the past landscape of these sites.

Aristotle was among the first to describe environmental changes in the Argive area. In the Meteorologica, Aristotle stated that in the Late Bronze Age the area around Argos was swampy while Mycenae had excellent soil. By Aristotle's time (4th century B.C.E.), the land around Mycenae had become too dry for agriculture, but the land of Argos was arable. According to Curtius (1851), by Roman times, Pausanias (2nd century C.E.) described the Lake of Alcyon in the wetlands near Lerna as a "bottomless" lake for which Nero endeavoured, in vain, to ascertain its depth. This lake persisted into the 19th century. Other early travellers, Leake (1830) and Gell (1840), deduced that the sea must once have been as far inland as Tiryns.

In the nineteenth century, the French "Expedition scientifique de Morée" (Boblaye & Virlet 1883) took a core approximately 200 m inland from the beach. Marine gastropod shells were found in a deposit below the modern floodplain sediments. This was the earliest scientific documentation for the progradation of the coastline. Later authors often cited this data as evidence that the sea once extended farther inland (Schliemann 1885; Philippson 1892; Maull 1921). Further evidence for a transgression of the sea came from Lehmann (1937) who calculated, from the slope of the coastal plain and the lowest excavated occupational layer at Tiryns, that the coast may have been only 350 m from Tiryns. More recently, Kraft and others (1977) studied four cores drilled between Tiryns and the shore. Based on this data, they postulated an extensive middle Holocene swamp along the present coast.

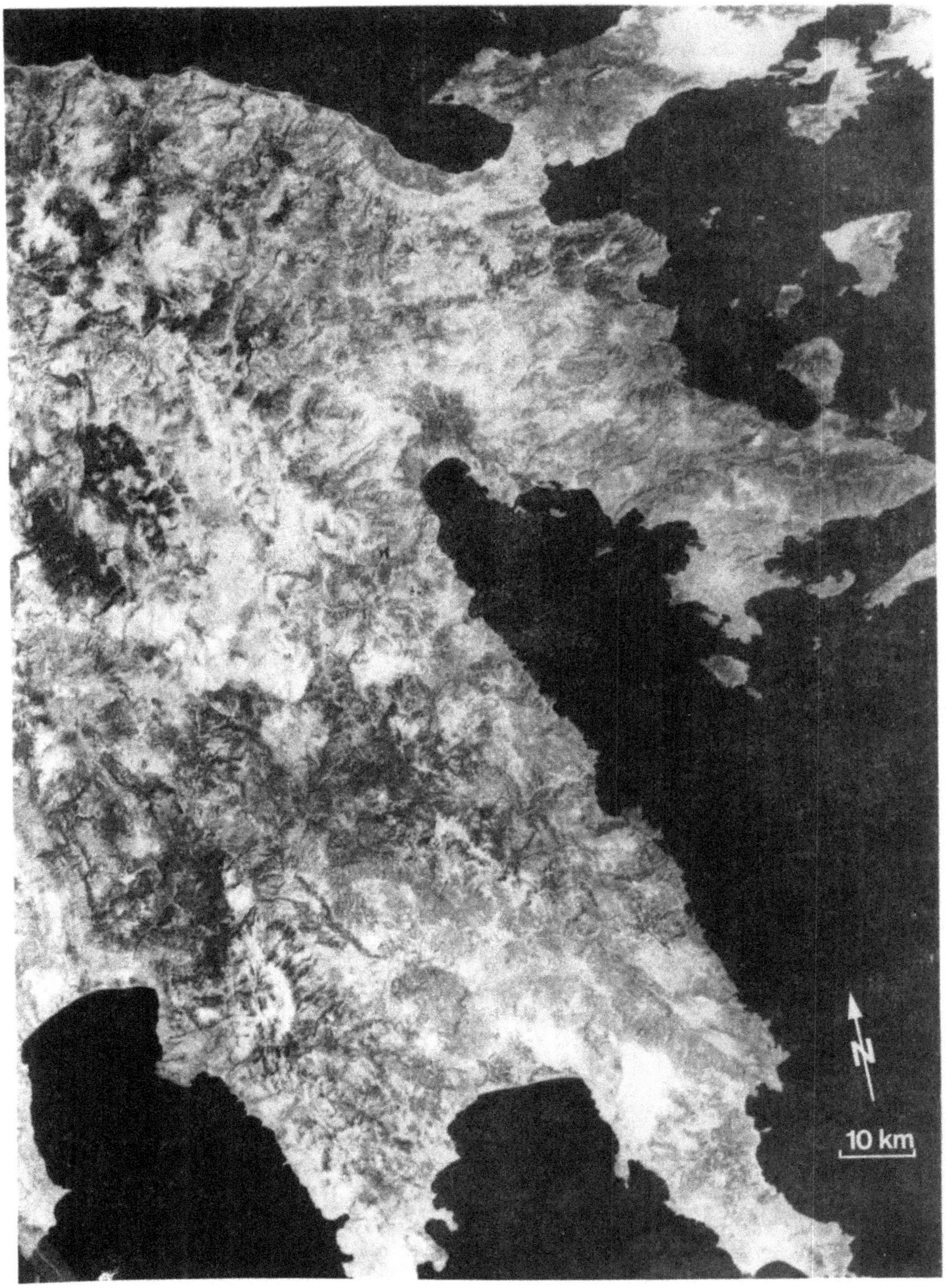

Figure 2: A photograph of a portion of a Landsat-1 image of the Peloponnese, Greece from August 2, 1972. The image is produced from reflectance information from MSS channel 5 (red wavelengths).

METHODS OF REMOTE SENSING

We will now examine the Landsat-1 coverage of the Argive plain and address whether units observed on the satellite image can provide specific information about the paleogeography of this coastal area. A section of the November 18, 1972 Landsat-1 image of the Peloponnese, Greece (scene identification #1118-08384) is used here. The area covered by one Landsat scene is approximately 185 km by 185 km; a standard photographic product of this scene is at an approximate scale of 1:1,000,000. Rotation of the earth during imaging creates a diagonal distortion in the satellite image. Figure 2 is a portion of a typical single band Landsat photographic scene of the Peloponnese from August 2, 1972 (scene identification #1010-08375).

The Landsat satellite system is equipped with a multispectral scanner (MSS) and spectral data of the Earth's surface is collected by the scanner from an orbit of about 900 km. The Landsat scanner has a spatial resolution that corresponds to a ground cell (called a pixel) of approximately 79 metres on a side. The data are recorded in four wavelength bands. Two bands (channels 4 and 5) are in the visible spectrum and correspond to 0.50-0.60 μm or green and 0.60-0.70 μm or red wavelengths. The remaining two bands, channels 6 and 7, are near infrared bands corresponding to 0.60-0.70 μm and 0.80-1.1 μm wavelengths. For additional information on Landsat instrumentation and spectral scanners, the reader is referred to Lillesand and Kiefer (1979) and Sabins (1987).

The Landsat-1 image used in this study is in a digital computer-compatible tape format (CCT). The data were analyzed using two approaches. First, we compared the visible features on the Landsat image with the known geology and vegetative cover to define geomorphic units. We then analyzed the spectral data from these units and produced a computer-classified land cover map which distinguishes areas with similar surface cover. Interpretation of the satellite data was done on a coloured computer display of the image, but for logistical reasons, are here presented in black and white.

THE ARGIVE PLAIN AS SEEN ON LANDSAT

The first level of interpretation of the data involves standard photogeologic interpretation of the satellite image. Four general geomorphic units (the bedrock, the alluvial fans, the floodplain, and the coastal area) and the bay waters are mapped on a reproduction of the Argive plain image (Figure 3a and 3b). There are clouds in the northeastern portion over the Arachnaion mountains.

UNIT 1 The bedrock consists of Triassic and Upper Cretaceous limestones as well as Tertiary marls and siliclastic flysch deposits. On the satellite image, this unit appears as bright regions with dark shadows on the northwestern face of the mountains creating a sense of the vertical relief.

UNIT 2 The alluvial fans, mostly Pleistocene in age, are present between the vegetation-rich floodplains and the barren and mountainous bedrock. The fans are mainly covered with olive trees and appear grey on MSS channel 5.

UNIT 3 The central portion of the image is the location of the Holocene floodplain deposits. These are planted with apricot and orange orchards. The November date of the image accounts for the strong infrared response (very dark on Figure 3) of the winter wet season growth.

UNIT 4 An asymmetrically-shaped coastal zone, characterized by a few artichoke fields and marshlands, separates the bay from the orchards. This zone varies texturally on the image from bright to dark.

UNIT 5 The gulf water is black on the Landsat image.

The harbour town Nafplion in the southeast, the town of Nea Kios on the shore at the mouth of the Inachos River, the modern city of Argos in the west central plain, and several smaller villages are also distinguishable on the image (Figure 3).

CLASSIFICATION OF THE IMAGE

The digital format of the image permits a computer analysis of the spectral data. This process involves choosing small areas on the Argive plain image to represent a distinct surface cover. The spectral data from the cover types are entered into a computer program which identifies other areas on the image with similar spectral reflectance characteristics. A description of the method involved in the computer classification method is given below.

A group of pixels, representing a known ground cover, is first delineated on the image. On the Argive plain image, seven cover types (Table 1) were defined on the basis of: vegetation density, land use, soil properties, and water variations. The reflectance information for each cover type is stored as a training set file. Next, a statistics program computes the mean and standard deviation of each of the wavelength bands in the cover type training sets. Cover types with large standard deviations indicate a greater variability in the reflectance properties. A computer classification program compares the four mean digital numbers (proportional to reflectance values) for each of the cover types to the image pixel value. If the image pixel lies within two standard deviations of the mean values of a cover type, it is assigned to that cover type. Pixels not within two standard deviations of the means for any of the cover types remain unclassified. The training set of each cover type must contain enough pixels to represent the spectral characteristics of the group but be limited

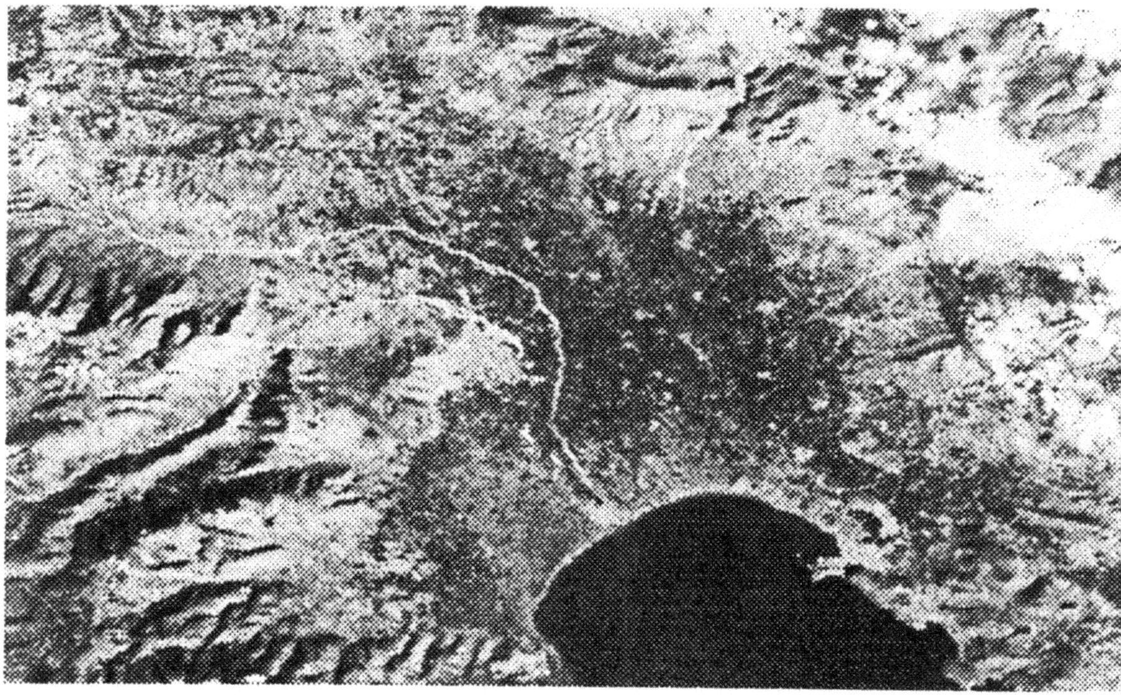

a)

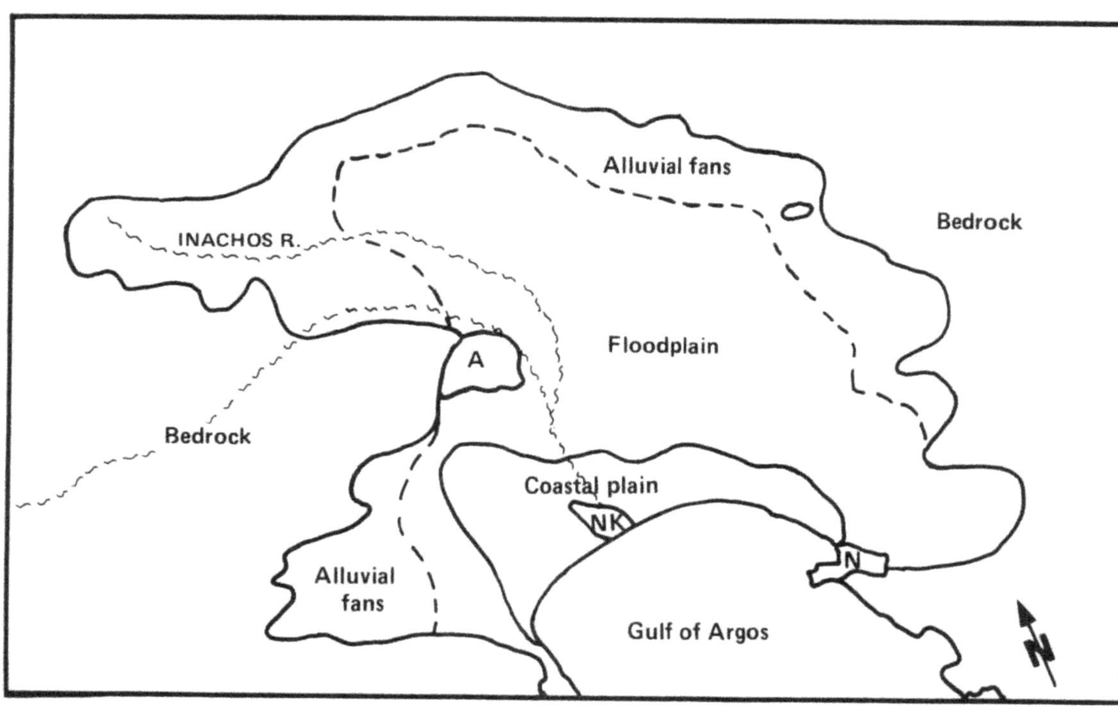

b)

Figure 3: a) A portion of the Landsat-1 image from November 18, 1972 of the Argive plain, Greece used in this study. The image is here displayed in MSS channel 5. b) Diagrammatic map of the above portion of the satellite image identifying five geomorphic and geographic units. The modern cities of Argos (A), Napflion (N), and Nea Kios (NK) are marked.

enough in size to reduce large variances which might overlap other groups.

A general spectral reflectance curve (Figure 4) is a plot of the multispectral scanner (MSS) channel against the relative reflectance value for each cover type and graphically represents similarities between the groups. The graph illustrates that three major groups have distinct spectral properties -- vegetated, non- to sparsely-vegetated, and water. MSS channel 5 is the wavelength band with the maximum separation of cover types.

Table I: Cover Types used in Classification

Cover type	Description
1	Alluvial sediments
2	Dense vegetation
3	High density vegetation
4	Coastal vegetation
5	Shallow water
6	Deeper Gulf water
7	Urban areas

Cover types 2, 3 and 4 exhibit characteristic vegetation spectral reflectance -- high green and near infrared (NIR) reflection values and low red wavelengths values due to the absorption of energy by chlorophyll within the plants. Variation in relative reflectance magnitude distinguishes the coastal vegetation cover type (4) from the other vegetation. A lower NIR response in the coastal vegetation may be due to an increased amount of water retention in this coastal setting.

Bright and dry surfaces of the non- to sparsely-vegetated alluvial deposits and urbanized areas (cover types 1 and 7, respectively) have a high reflectance in the green and red wavelength bands. The values in the urban cover type have relatively stronger responses possibly due to smooth concrete walls and roofs of the buildings creating large bright surfaces which increase reflectance.

Water has a characteristic low near infrared reflectance and its presence in any of the cover types will effectively lower the NIR value. The shallower nearshore water has a higher reflectance value in each of the band wavelengths than the deeper water. This is probably due to either a brownish colour of the suspended sediments in the water or to detection of the bottom in very shallow water.

RESULTS OF REMOTE SENSING TECHNIQUES

The outcome of the spectral analysis is a computer-classified land cover map (Figure 5) which emphasises areas with similar spectral reflectance. The spectral data varies according to the soil, vegetation,

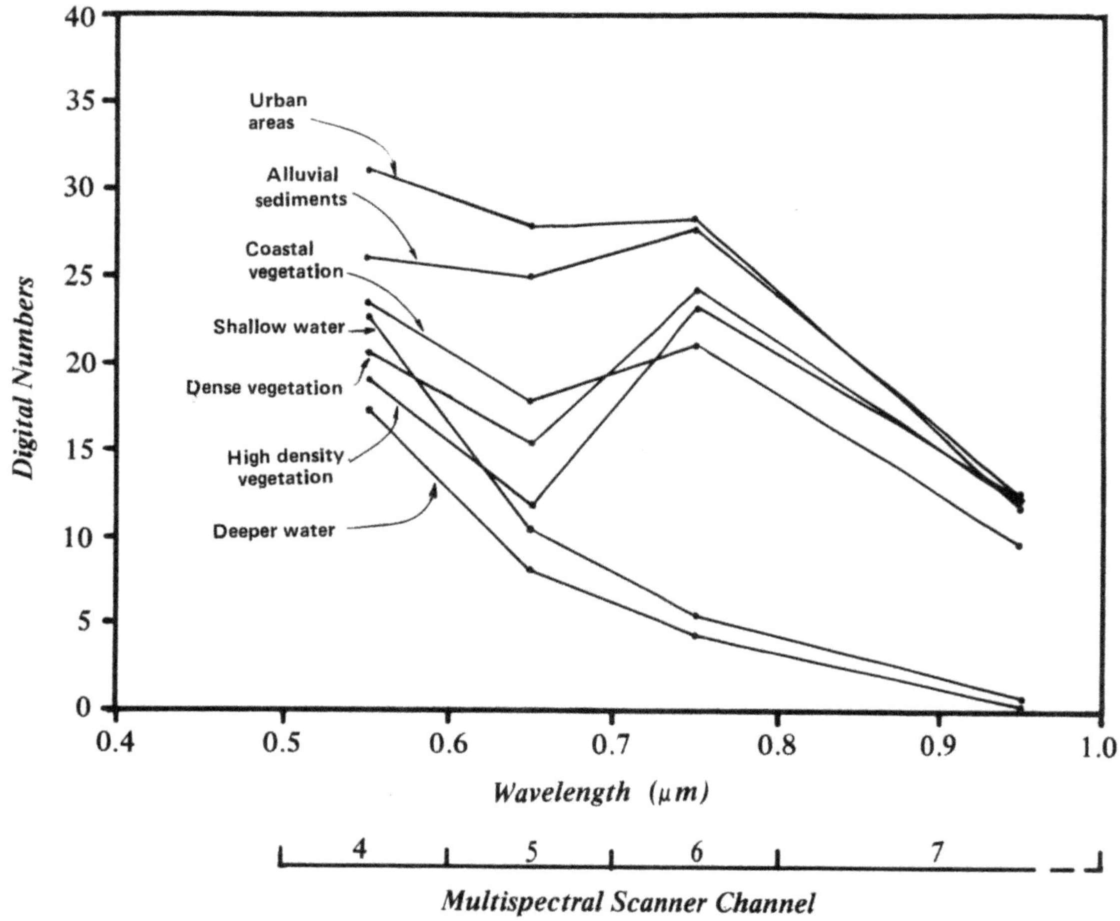

Figure 4: Relative spectral reflectance curve is a plot of the mean digital numbers (proportional to the reflectance values) for each of the cover types against the median wavelength of each of the MSS channels of the Landsat-1 satellite.

water content and other factors controlling the surficial cover. The main geomorphic features identifiable on the computer-generated land cover map are: 1) the boundary between the alluvial fans and the floodplain of the central valley, 2) the boundary between alluvial plain and coastal plain, 3) the various vegetated and non-vegetated zones within the coastal zone, and 4) the characteristics of the nearshore waters.

The non- to sparsely-vegetated alluvial fans which predominantly flank the mountains east of the Argive plain are classified as alluvial sediments. The densely vegetated valley has a strong vegetation spectral reflectance which allows the central floodplain to be distinguished from the alluvial fans. There is also a rather sharp boundary between the agriculturally-rich floodplain and the coastal plain. A change of vegetation across this boundary enables us to detect it on the satellite image. The crescent-shaped coastal plain is characterized by an outer zone of vegetation-poor alluvial deposits and an inner area of coastal vegetation. The coastal vegetation is distinct enough in spectral properties to allow its separation from the other vegetation groups. In the centre of the coastal vegetation is a zone of dense vegetation. A lush growth of reeds in this location allows for its distinction from the other coastal vegetation.

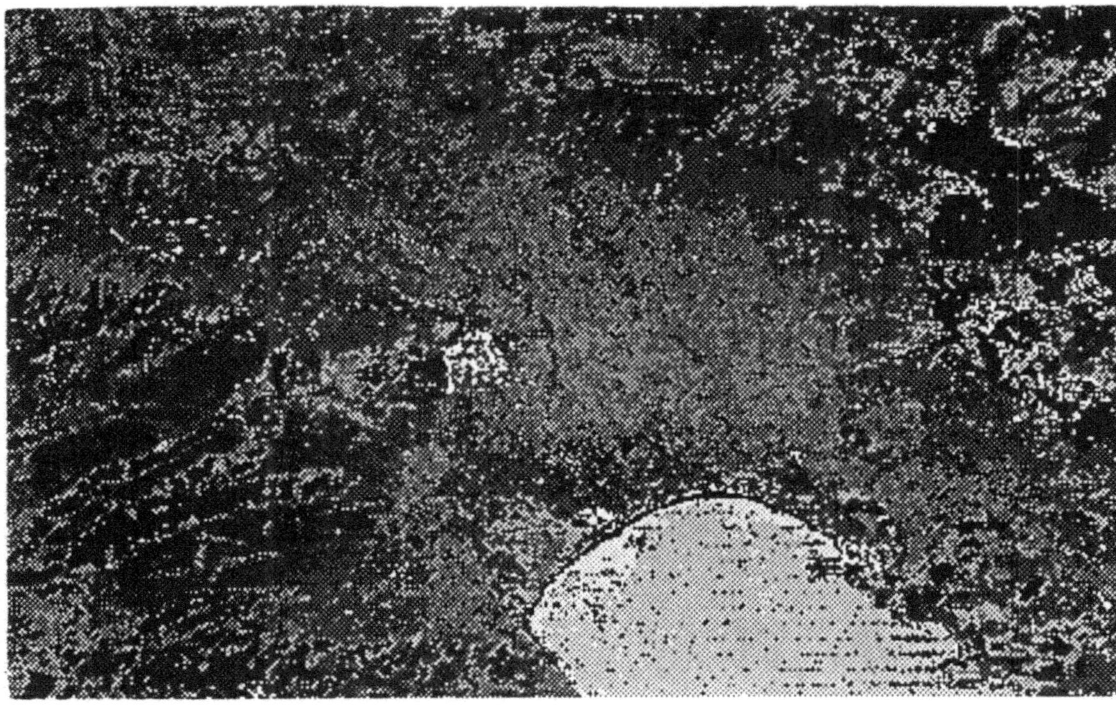

a)

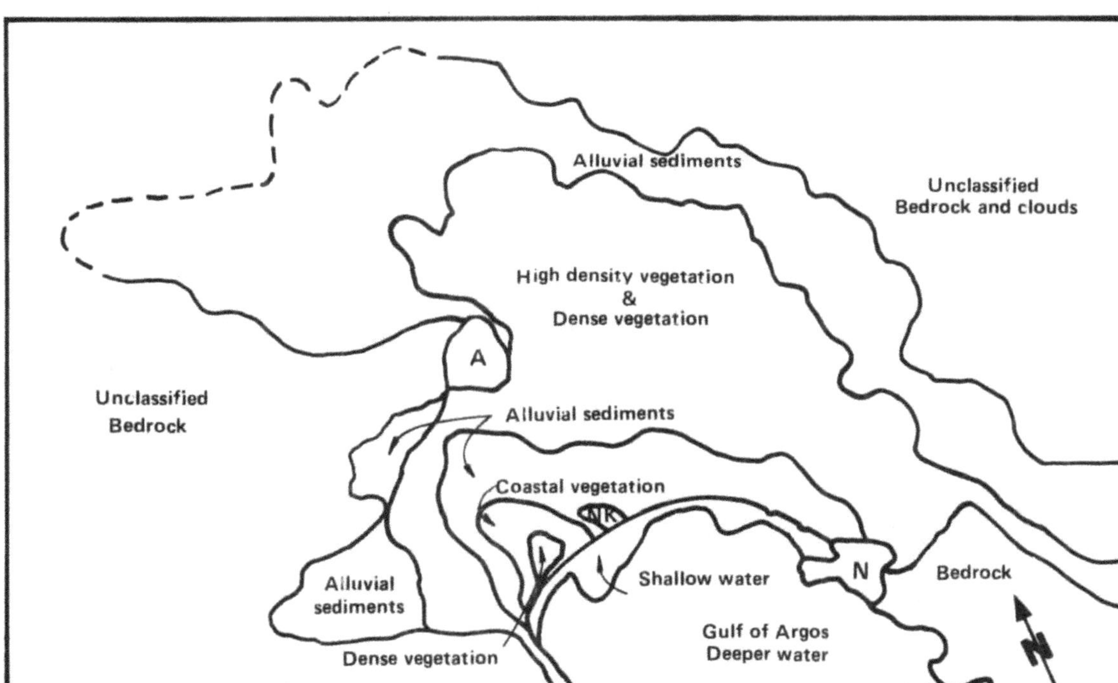

b)

Figure 5: a) A computer-classified land cover map of the Argive plain showing the distribution of cover types. Seven cover types were chosen on the basis of vegetation density, land use, soil properties, and water variations. b) Sketch map of the computer- generated map. Cover types are labelled.

Along the present shoreline is an unclassified area (black) which represents modern beach deposits. Just offshore, the distribution of cover type 5 (shallow water) indicates a concentration of shallow or muddier waters along the shoreline. The lobate shallow water area at the mouth of the Inachos River is a submerged delta front. The direction of longshore transport is seen on the computer classified land cover map in the westward skewing of the deltaic deposits.

Do the boundaries observed provide specific information about the Holocene paleogeography? To test our observations and methods of remote sensing, we compared the results of the image interpretation to subsurface core data.

SUBSURFACE INVESTIGATION OF COASTAL SEDIMENTS

METHODS

Subsurface data presented in this paper are based on 11 drill sites. Nine cores to a depth range of 4.1 to 10 metres were taken with a 10 cm diameter Eijkelkamp hand auger. Two 30 cm diameter holes (17 and 20 m deep) were drilled using a bailer rig by a commercial drilling company. The auger and drill sites are aligned in one cross section stretching 5.0 km from the margin of the town of Argos to the Gulf of Argos. In addition to the 11 core sites aligned in cross section, stratigraphic information from 14 other core sites is compiled here (Figure 6).

Auger cores were described in the field. Descriptions include: lithology, colour, soil properties, and macrofossil content. Samples were taken for laboratory analysis of soil properties, grain size, microfossil identification, and dating. Samples from the bailer drilling were taken at one metre intervals and sieved in the field. Mesh widths of 0.063 and 2.0 mm were used to split the sample into sand and gravel fractions. The sand fraction was analyzed using a binocular microscope to recover microfossils and to describe shape, composition and texture of the grains.

Potsherds found in the cores were used to assign maximum ages to the strata. These ages were determined by archaeologists of the Deutsches Archäologisches Institut in Athens. In addition, one radiocarbon date was obtained from a peat in core AP 49 at a depth of 5.8 metres. Correlation between cores is based on radiocarbon date, relative ages determined from soil development, archaeological dates, and sedimentary units which were used as marker horizons.

STRATIGRAPHY

The lowest sedimentary layer (Unit A, Figure 7) is a sequence of consolidated , red soils developed in alluvial sediments. The well developed soil properties -- thick clay films and calcium carbonate nodules >1 cm in diameter -- indicate that this unit is of Pleistocene

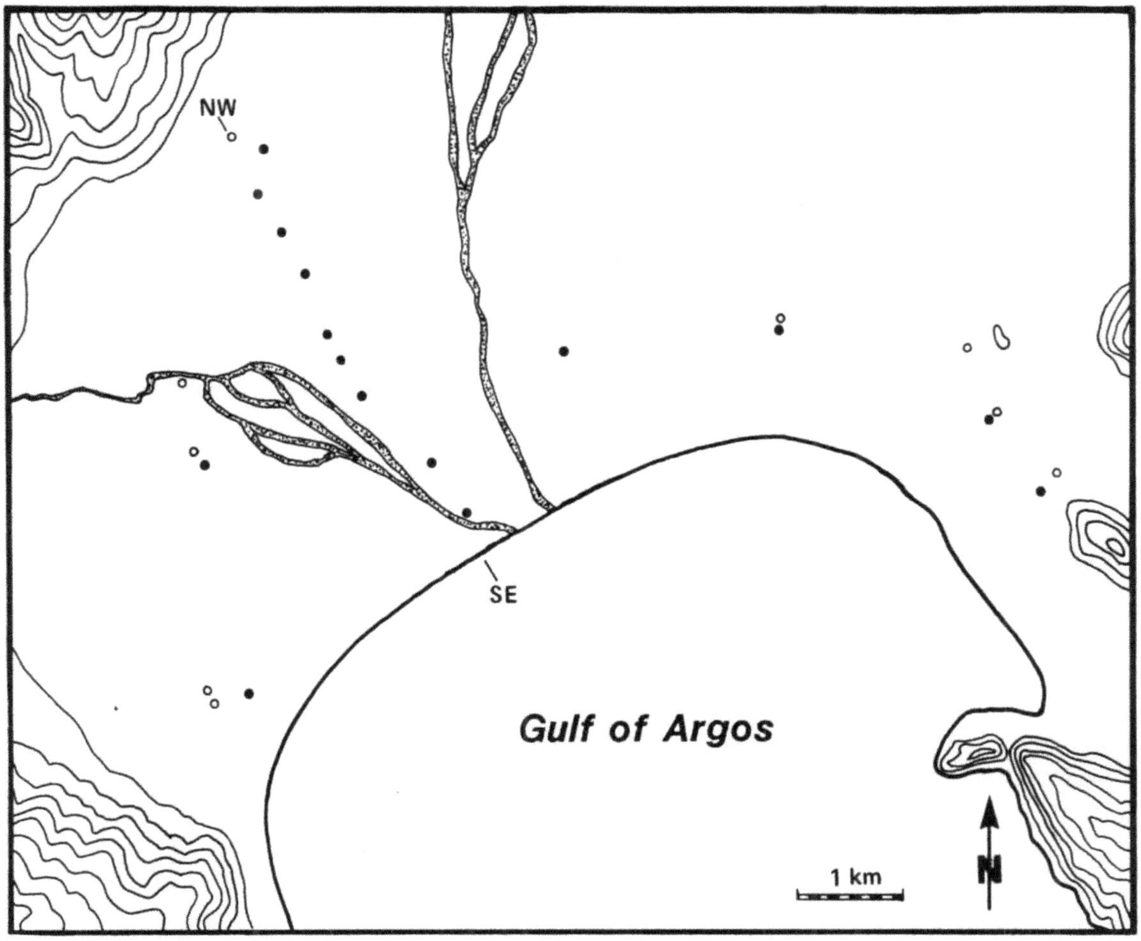

Figure 6: The Argive plain coastal zone with core location. Transgressive deposits are present in sites marked with filled circles, and are absent in sites marked with open circles. The stratigraphic sequence of the NW-SE line of section is described in this paper. Contour interval = 50 m.

age. Locally, the Pleistocene soil unit is overlain by a younger soil (Unit B). Unit B lacks the nodular carbonate, advanced clay films, and red coloration found in the older soil, and it does not contain potsherds. This soil is therefore inferred to be early Holocene in age.

Blue-grey silts and clays (Unit C), deposited in a marine to lagoonal environment, bury the early Holocene and Pleistocene sediments. These Unit C muds thin landward, reaching a maximum thickness of six metres near the shore, and disappear about 4.8 kilometres inland. Peat beds which include both brackish and freshwater shells, are intercalated with the muds. One peat sample (AP 49-5.8) from approximately 3 metres above the late Pleistocene-early Holocene soil and 900 metres inland, yielded a corrected radiocarbon age of 3820 ± 100 B.C. (calibrated after Pearson et al., 1986). The distribution of the Unit C facies can be mapped using stratigraphic data from cores off of the line of section (Figure 6). Unit C is characterized by a landward thinning and a definable inland extent.

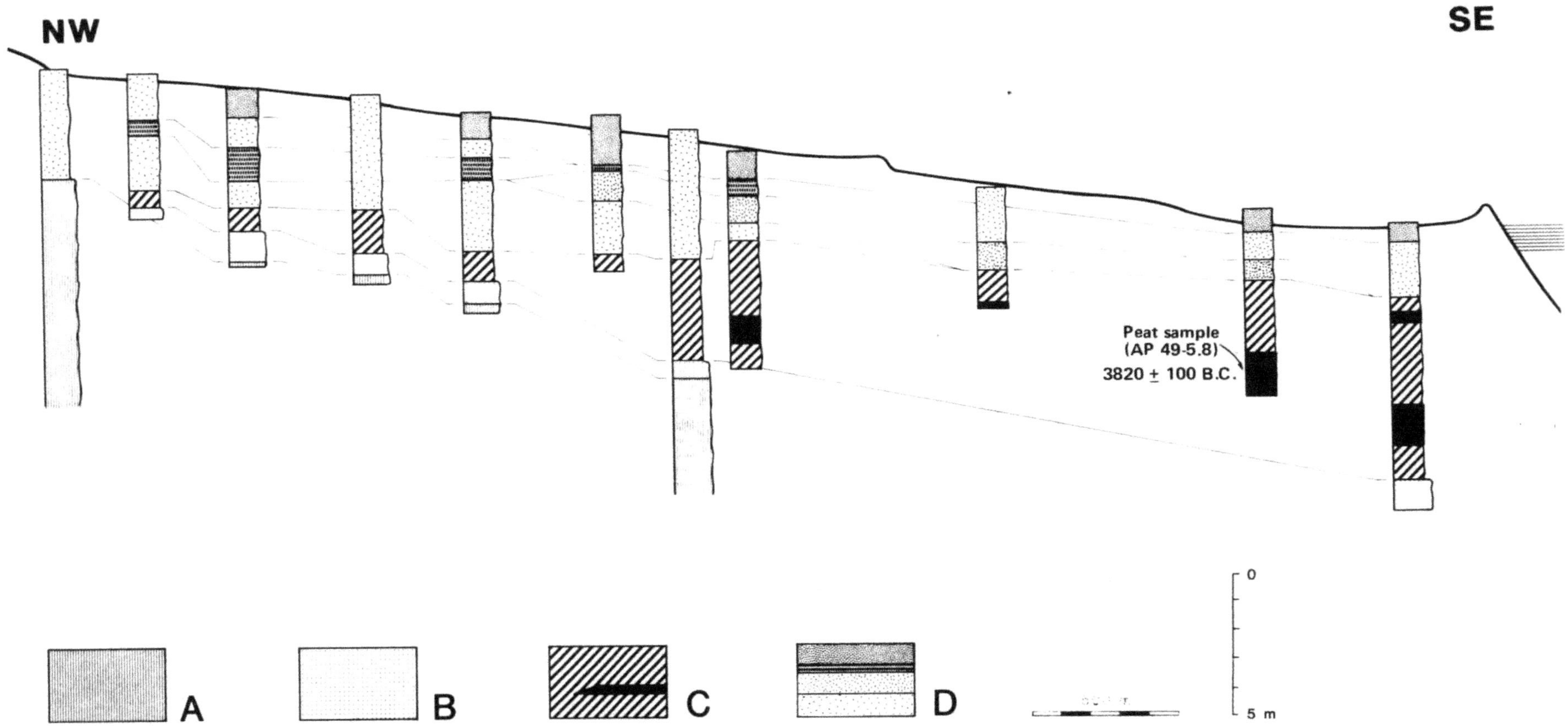

<u>Figure 7:</u> Cross section of drill core data with designated stratigraphic units. Unit A is a red Pleistocene soil sequence with calcium carbonate nodules greater than 1 cm in diameter. Unit B is an early Holocene soil lacking advanced carbonate and clay film development. Both soils indicate periods of land stability.
The deposits of Unit C are blue-grey clays and silts with some peat beds marked in black. Unit C muds are interpreted as subaqueous deposits and its inland extent marks a maximum transgression. Unit D consists of interbedded alluvium and soil up to 4 metres thick. Cores above the line of section represent core sites off of the line of profile at a slightly different elevation.

The deposits of Unit D, yellowish brown clay loam divided into separate horizons by soil development, overlies Unit C. Potsherds found within Unit D range from Neolithic to Hellenistic age. More than 2 m of post-Hellenistic alluvium is present in the upper section of this unit.

DEPOSITIONAL HISTORY

Unit C muds record an early to middle Holocene marine transgression which began before 3820 ± 100 B.C. since the dated peat (AP 49-5.8) was located well above the base of Unit C. The inland limit of the subaqueous Unit C deposits marks the maximum shoreline of this

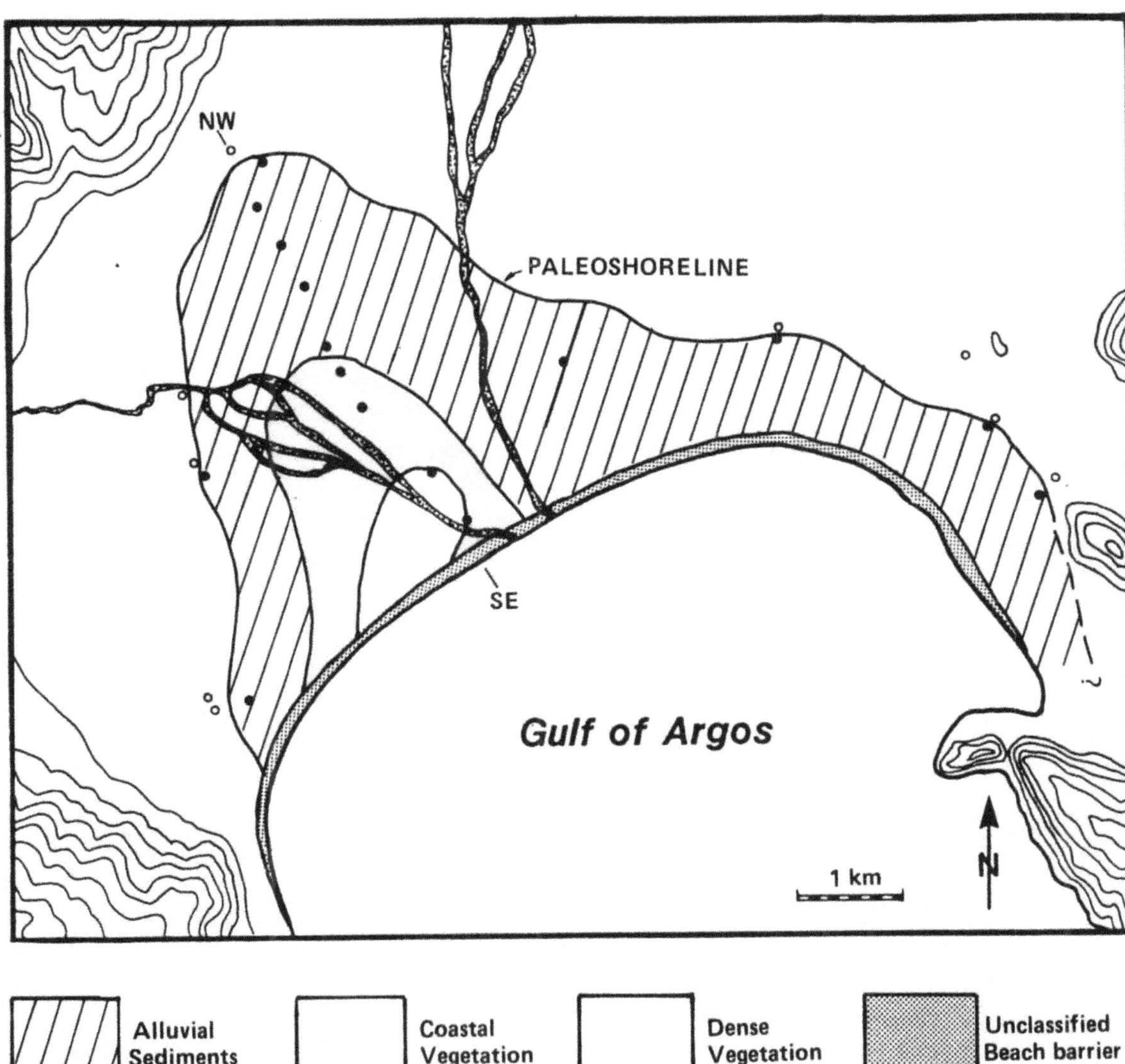

Figure 8: Projection of boundaries seen in the coastal region of the Argive plain on the computer-classified land cover map onto location map of subsurface core sites marking the presence of transgressive muds (solid circles) and their absence (open circles). The boundary between the alluvial sediment-covered coastal area and the densely vegetated floodplain seen on the satellite image correlates to the inland limit of the subsurface marine to lagoonal mud unit, marking the maximum early to middle Holocene transgressive shoreline.

transgression. The palaeoshoreline corresponding to the transgression is the boundary between cores with Unit C muds and cores where it is absent (Figure 6).

Landward of a beach barrier which developed during the transgression, peats were deposited in a lagoon. Snail shells indicate that brackish to freshwater conditions prevailed due to the discharge of river as well as seawater into the lagoon. Although most of the lagoon has been filled with sediment and is now overlain by floodplain alluvium of Unit D, in Nea Kios remnants of the wetland conditions persist. During shoreline progradation periods of land stability and soil formation were separated by periods of sediment aggradation.

We turn now to the comparison of the subsurface stratigraphy and the facies distribution with the Landsat imagery.

CONCLUSIONS

Geomorphic units identified on the Landsat-1 image and computer classification methods of remote sensing show the areal distribution of various surficial cover that are defined on the basis of vegetation, land use, soil properties, and water variations. Figure 8 is a map of the coastal zone of the Argive plain which illustrates a projection of computer-classified cover types from the satellite image onto the location map of the core sites. From the comparison of the two data sets (satellite and subsurface), we see that the sharp boundary between the agricultural-rich alluvial valley and the coastal plain seen on the satellite image as a vegetation change correlates to the inland limit of the subsurface Unit C marine to lagoonal muds. The spectral reflectance of coastal plain at the boundary to the floodplain is characterized by the non- to sparsely-vegetated alluvial sediment cover type. Since we know from the subsurface data that the Unit C muds do not extend farther inland under the sediments of the central valley, we infer that the subsurface transgressive muds affect the surface geology (i.e. change in soil, vegetation, or water properties) enough to allow the detection and the identification of a palaeoshoreline.

Within the western side of the coastal plain, the area mapped as coastal vegetation (Cover type 4) correlates with the subsurface peat deposits from the line of section. This is the probable location of a middle Holocene lagoon. At the centre of the Holocene deposits, the densely vegetated (Cover type 2) area correlates to the modern day wetlands which are remnant of an earlier lake.

Of particular interest to coastline reconstruction studies, the maximum extent of the early to middle Holocene transgression is recorded in the satellite image even though the deposits are covered by 4 metres of modern floodplain deposits. Since Landsat satellites record only the reflectance properties of the upper 0.01 mm of the surface, we can deduce that the surface variations seen as vegetation, soil, and water reflectance changes on the Landsat image are controlled by the surficial geology which can be correlated to subsurface units. Interpretations of the Landsat data as presented here, however, must be verified by fieldwork. This method of satellite data interpretation is not meant as a replacement of fieldwork but as a reconnaissance technique to minimize the amount of necessary field sampling and to maximize the area of potential study.

ACKNOWLEDGEMENTS

We wish to thank the Stanford Remote Sensing Lab and R.J.P. Lyon for supplying the Landsat-1 CCT used in this study and for the generous use of equipment, programs (written by Kai Lanz), and supplies. The results presented in this paper are part of the "Argive Plain Project", a cooperative investigation carried out by the "Deutsches Archäologischen Institut" in Athens and the Department of Geology at Stanford. We would like to thank the German School for their logistic help in the field and Tjeerd van Andel for his supervision of the project. Dr. H. Haas at the radiocarbon laboratory of the Institute for the Study of Earth and Man in Dallas provided dates for peat sample AP 49-5.8. Thanks are extended to Lisa E. Wells, Kevin O. Pope, Andy Tomlinson and others who reviewed drafts of this paper and to Cleavy McKnight for assistance in production of the image figures.

REFERENCES

Baeteman, C. 1985. Late Holocene Geology of the Marathon Plain (Greece). Journal of Coastal Research 1 (2): 173-185.

Boblaye, E.P. and T. Virlet. 1883. Expédition Scientifique de Morée. Section des sciences physiques II, 1 Géographie, 2 Géologie.

Curtius. E. 1851. Peloponnes - eine Historisch-Geographische Beschreibung der Halbinsel. Volume 1: 495.

Finke, E. 1984. Die Entwicklung der Bucht von Volos im Holozän. Unpublished master's thesis, Kiel University: 150p.

Gell, W. 1810. Argoli.

Gifford, J.A. 1985. Post Bronze Age Coastal Change in the Vicinity of Kition. In: Excavations at Kition. V. Karageorghis and Demas (Eds.).

Kraft, J. 1972. A Reconnaissance of the Geology of the Sandy Coastal Areas of Eastern Greece and the Peloponnese. Technical Report No. 9, College of Marine Studies, University of Delaware, 160p.

Kraft, J., S.E. Aschenbrenner and G. Rapp. 1977. Palaeogeographic Reconstructions of Coastal Aegean Archaeological Sites. Science 195: 941-947.

Leake, W.M. 1830. Travels in the Morea. Murray, London.

Lehmann, H. 1937. Argolis - Landeskunde der Ebene von Argos und ihrer Randgebiete. Herausgegeben vom Deutschen Archäologischen Institut, Athen: 150p.

Lillesand, T.M., R.W. Kiefer. 1979. Remote Sensing and Image Interpretation. John Wiley & Sons, New York. 612p.

Maull, O. 1921. Beiträge zur Morphologie des Peloponnes und des Südlichen Mittelgriechenlands. B.G. Teubner Verlag: 120p.

Pearson, G.W., J.R. Pilcher, M.G.L. Baillie, D.M. Corbett and F. Qua. 1986. High-Precision C14 Measurement of Irish Oaks to Show the Natural C14 Variations from AD 1840 to 5210 BC. Radiocarbon 28: 911-934.

Phillippson, A. 1892. Der Peloponnes - Versuch einer Landeskunde auf Geologischer Grundlage. Frieslander, Berlin.

Sabins, F.F. 1987. Remote Sensing, Principles and Interpretation, 2nd ed. W.H. Freeman & Co, San Francisco: 449p.

Schliemann, H. 1885. Tiryns - the Prehistoric Palace of the Kings of Tiryns. J. Murray, London - Reissued in 1967 by Benjamin Blom, Inc. New York 10452: 385p.

MARINELLA PASQUINUCCI, GABRIELLA ROSSETTI

The Harbour Infrastructure at Pisa and Porto Pisano from Ancient Times until the Middle Ages

ABSTRACT

From ancient times until the Middle Ages Pisa used several ports and landing places, in relation to the dimensions and cargos of ships and to the evolution of the coastal morphology. The most important of them, called Portus Pisanus/Porto Pisano and located north of Livorno, about 16 km from Pisa, is studied on the ground of archaeological, literary, epigraphical and documentary evidence.

ANCIENT TIMES

As archaeological-topographical research and examination of mediaeval and modern documentation show, the coastline at the point of the Arno delta in the 2nd and 1st centuries B.C.E. was about 7 km further inland than it is today. Subsequently, there was a gradual and general advance, probably interrupted now and then for brief intervals and even possibly receding slightly; the process continued up to about the middle of the 19th century. Following this point the phenomenon was no longer all-embracing; but small, general withdrawals and considerable, partial withdrawals took place, especially at the mouth of Arno (Mazzanti and Pasquinucci 1983a; 1983b, in print). (Fig. 1).

From the end of the 1st century B.C.E., huge quantities of sediment were carried down to the sea by the Arno (especially as a result of deforestation carried out by the centurations and the increase in the number of agricultural settlements). These quantities were such as to counteract the effect that a possible rise in sea-level would have had in producing a withdrawal of the coastline; they were also such as to cause the delta to advance by about 7 km in little more than 2000 years, together with the sandy beaches on each side of it (stretching beyond Viareggio, together with the River Serchio, and as far as Livorno).

These changes in the coastal strip, and man's subsequent presence there, have made it difficult to work out exactly where the former coastline was, in particular, in the area north of Livorno where, as we shall see, the Portus Pisanus was sited. This has led to various contrasting hypotheses being put forward as to its location (Toscanelli 1933-34; Schmiedt 1964; 1978; cfr. Banti 1943).

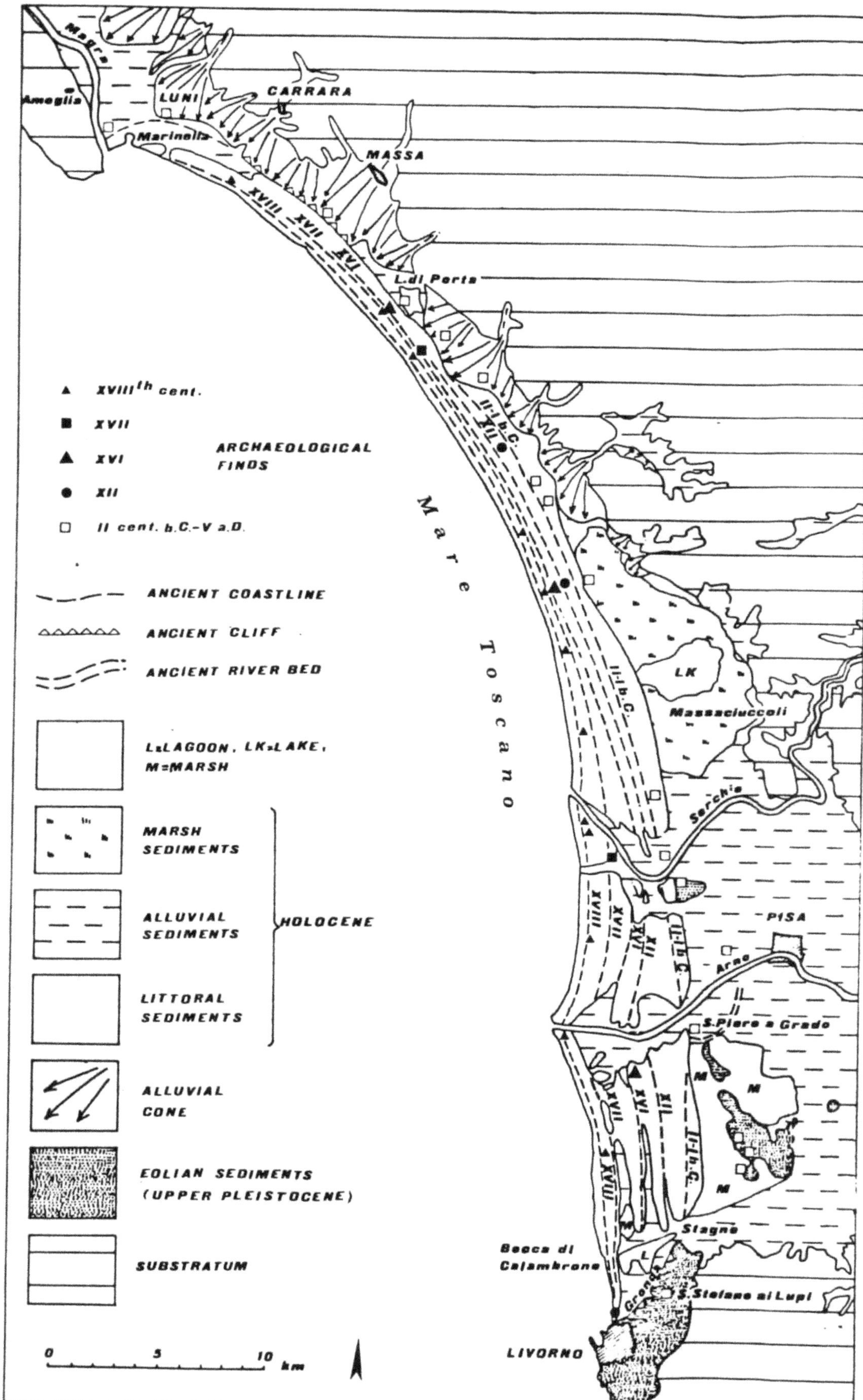

Figure 1: The evolution of the Luni-Pisa coastline (II-I cent. B.C.E. XIX cent.) (dr. R. Mazzanti after Inghirami map, 1830).

Geomorphological investigation including core data closely coupled with archaeological-topographical research, now makes a reappraisal of the question possible, using new lines of argument, as well as adding further information to our knowledge of the Portus Pisanus.

It is possible to hypothesize the existence of various landing places, in the pre-Roman era, located:

1. in the city;
2. along the river Arno (certainly at San Piero a Grado, where there was a series of settlements, attested to from the Late Iron Age to the Imperial era, which played a fundamental part in the linking of the coast and the city, both overland and by water. It should be noted that these settlements are near the church, which is connected with the tradition of Saint Peter having landed there, and that, in the 1st to the 3rd centuries A.D., this was the site of a large set of buildings, made up of at least 12 practically square rooms arranged in two groups of 6 on the sides of a larger rectangle, which may possibly be taken to be the structure of a port or a villa (Pasquinucci 1986);
3. along the coast, to the north and to the south of the Arno. Among these landing places there would seem to have been what sources dating back to the Imperial era already call Portus Pisanus (Pasquinucci 1986; Ciampoltrini, Cianferoni, Romualdi 1982-83).

In the late Republican and early Imperial eras, literary sources talk of Pisa and not of a separate port, even in reference to the fleet (Lugand 1926; Banti 1943). In the Imperial era, Strabo mentions activity to do with ship-building (5,2,5) and, in the 4th century, Claudian (bell. gild. 483) refers to the shipyards (navalia) of Pisa (which, going by a conjecture which is impossible to prove, were situated at the far western end of the south bank of the Arno, in the region of what was called the Terzana in the Middle Ages (Schmiedt 1978); guilds of fabri navales and fabri tignarii are attested in CIL XI I 1436 = ILS 7258, whose attribution to Pisa there seems no reason to doubt (Banti 1943; Ciampoltrini 1981).

Literary sources record that a Portus Pisanus, distinct from the city of Pisa, existed at the beginning of the 3rd century A.D. (Itin. Marit. 501). It may even have been in use earlier. It certainly must have become more important in the Imperial era when, together with the increase in the tonnage of ships the Arno must have become increasingly difficult to ascend by ship because of the increase in the amount of sediment already mentioned (see Strabo 5.2.5, whose source already points out how difficult navigation from the mouth upstream was).

That Portus Pisanus was located in the area north of Livorno, between the present-day mouth of the Calambrone river and the "Gronda dei Lupi", as numerous scholars have already maintained, is beyond doubt. The "Gronda dei Lupi" is an escarpment which, just north of Livorno and stretching to the farm at Suese, separates the Holocene plane from the upper Pleistocene terrace on which the city is built. In the second and first centuries B.C.E., the coastline ran along the foot of the "Gronda", on a considerably more inland course than at present, probably, forming an inward loop in relation to the low, but rocky headland on which Livorno was later to be built (Mazzanti and Pasquinucci 1983a; 1983b in print; Finzi Contini, in print).

The stretch of water opposite this loop in the coastline was suitable for ships to moor in, both because it was relatively sheltered from the prevailing winds and protected by shallows and half-submerged bars, and because into it flowed the effluent (later called the Calambrone) of the marshes of Stagno. Unlike the treacherous mouth of the Arno, the Calambrone must always have been navigable, owing to the presence upstream of a large, very low-lying area, part of which was perpetually flooded and part of which was liable to flooding that weakened the surge of the periodic floods of the river (Mazzanti, Pasquinucci, Salghetti Drioli 1984; Mazzanti and Pasquinucci, in print) (Fig. 2).

The flood waters reached into the sea at the southernmost point of the sandy coast, along the "Gronda dei Lupi". The ebb-tide currents, with their movement over the increasingly large bars and banks forming to the west, which increase the extent of the lagoon to the rear, prevented the banks from joining up and allowed the larger lagoons, such as the Altura and the Galanchio, to remain sufficiently deep to be used as a harbour. It was no accident that mediaeval watch-towers were built along the course where these currents flowed.

The data obtained by means of geomorphological research agree with those of Itinerarium maritimum which gives Portus Pisanus as being 18 miles from Vada (the same name is used today) and 9 from the mouth of the Arno towards Pisa (Itin. Marit. 501).

During the Imperial era, the infrastructure of the harbour had to be steadily enlarged, Rutilius Namatianus (de red. 1, 531 ff) gave a lively description of the natural features of the site and of the bustling mercantile activity carried on there. Famous for its goods and its wealth of maritime traffic, the port astounded him (1, 531f.). It had direct access to the sea, and was protected by procera alga (1, 537: undoubtedly a bed of Posidonia covering the submerged bars) whose long, soft, ribbon-shaped leaves broke the impact of the waves and which gave the impression that the hulls were not being damaged (de red 1, 533-540 and 537-540 in particular). The text also supplies useful information concerning the coastal scenery: at Pisa Rutilius took "the usual overland route", and he was offered horses and even a coach for the journey (de red. 1, 560 f.); when a storm prevented him from continuing his journey by sea, he went hunting in the nearby woods (de red 1, 621-630).

Remains of buildings and tombs of various kinds are certainly to be connected with the Portus Pisanus, no doubt of the community existing there; these remains were located in the area between the Via Aurelia, the River Cigna and the fountain of Santo Stefano (Targioni Tozzetti 1768).

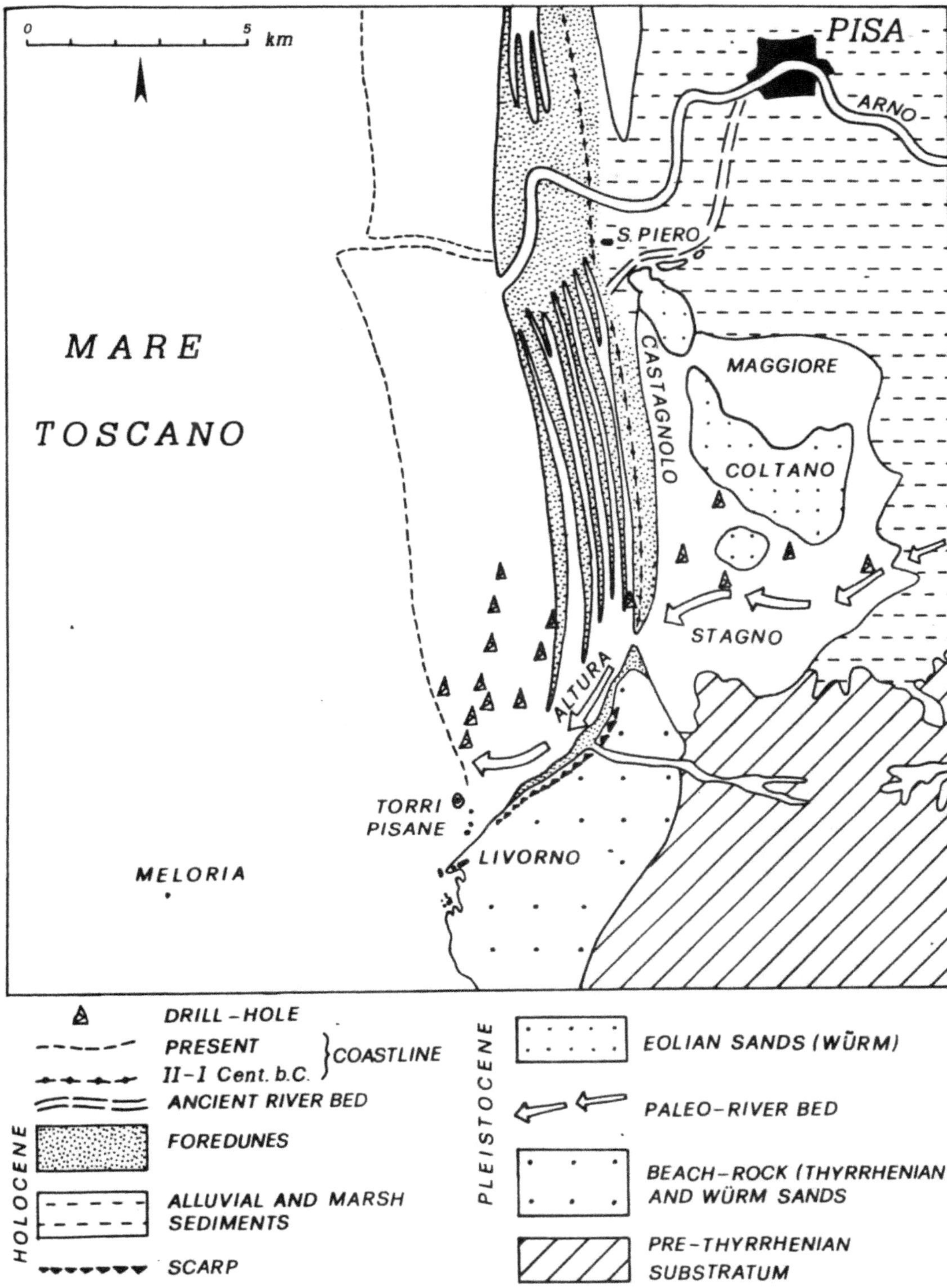

Figure 2: The Porto Pisano area in the XIIIth century.

Although no trace of these remains today we know of their existence from the description given by Targioni Tozzetti (1768), who saw the remains of the buildings and a large amount of Roman material visible at ground level, in 1742-1743, part of which he collected.

Another source is the account of excavation work carried out on the basis of the information supplied by the same Targioni (1768). Later, at the beginning of the 19th century, excavation and movement of earth took place on the site at the hands of owners of several of the farmsteads (Chiellini, Toccafondi, Sambaldi); very few details about this work have come down to us (Ciampoltrini, Cianferoni and Romualdi 1982-83).

Although the very nature of its retrieval and transmission to the present day does not provide an overall picture of the old community existing there, recent examination of the material identified as coming from the site in question, shows that the area was inhabited at least from the 3rd century B.C.E. and that it was very prosperous from the 2nd to the 4th centuries A.D. Finds dating from the beginning of the 5th century A.D. have also been reported. The prosperity of the port in the 4th century A.D. perhaps, partly due to the massive financial interests which certain great senatorial families of Rome had in it, is attested to by inscriptions that show, inter alia, evidence of close connections with the nearby port of Vada Volaterrana (Ciampoltrini, Cianferoni, Romualdi 1982-83).

Something which has not yet been made public is the recent discovery, during agricultural work, of an impressive set of remains. On the basis of finds at ground level, these may be reckoned to cover the period between the late Republican era and the 4th or 5th centuries A.D. Situated at Campacci, about 1.4 km (a milium) south-west of Santo Stefano ai Lupi, the set of buildings was probably near, or set alongside the coastal road leading from Vada Volaterrana, through Castiglioncello and Salviano, to Portus Pisanus.

Portus Pisanus and Pisae were linked both by land and by waterway. The road, or roads, between Portus Pisanus and Pisae, one of which is thought to have been coastal via San Piero a Grado, made it possible to go quickly from one place to the other (Pasquinucci 1986). Although this cannot be proved, today, there may also have been canals which might have been an integral part of the network of centuriation. Furthermore, it was possible to go up the Arno, with suitable craft and loads. As we have seen, there is evidence of a landing place in use from archaic times and of a settlement dating back to the Imperial era (Pasquinucci 1986) at San Piero a Grado, close to the point where the coastline ran in the 2nd and 1st centuries B.C.E.

A number of underwater finds that cannot be definitely attributed to any site or sites in particular, show that navigation was intense on the stretch of sea off Livorno from the 7th-6th century B.C.E. to the 3rd-4th century A.D. (Romualdi 1982). Material from the ager Pisanus makes it possible to trace archaic and late-archaic maritime trade routes from Etruria, from Greece proper, from Eastern Greece and from the Marseille area. Material also exists from central and southern Tyrrhenian Italy and from Attica, dating to between the end of the 5th and the 4th-3rd centuries B.C.E. In the late Republican and Imperial periods, sea trade brought goods to Pisa from Italy (especially from central and southern Tyrrhenian areas, but also from the Adriatic) and from elsewhere (Spain: Tarraconensis and Baetica; Gaul; North Africa: Proconsularis, Tripolitania). There is also evidence to show that Pisa had connections, by river, with the Arno valley and, by land, with Volterra and North Italy.

Archaeological finds from Pisa and its territory show with ever-increasing clearness, that Pisa was an emporium available to sea-going traffic through its various landing-places, the most important of which in the Imperial era, was Portus Pisanus.

THE MIDDLE AGES

At the beginning of the 7th century, Gregory the Great (Epist. 13, 33; Migne 77,1284) mentions the maritime activity of the people of Pisa but no explicit attestation exists of the use of Portus Pisanus at that time.

In the early Middle Ages, the name connected with the locality did not disappear, since the parish churches scattered around the area between Livorno and the Gronda dei Lupi are referred to as pleberii portus (Rossetti, in print). Probably, then, Porto Pisano was one of the ports on the Tyrrhenian sea that took on greater importance after the destruction of Luni in 860.

The mouth of the Arno was such that it was possible for galleys, narrow ships of limited draft, to pass; but already in the 11th century and at the beginning of the 12th, when Pisa's military exploits in the western Mediterranean were on the increase, serious problems had to be faced because the river-bed rose continuously and had to be constantly dredged. Describing the departure of the fleet for a venture off the Balearic islands (1113-114), the author of the Liber maiolichinus tells how the ships had to shed weight because of the difficulty in passing through the mouth of the river and how they waited for high tide to be able to take to sea, where they reloaded arms and supplies and sailed off (Calisse 1904, p. 13, vv. 163-169).

In the middle of the 12th century, in a busy decade of public works, the Commune fitted Porto Pisano out anew to meet the needs of the rapidly expanding trade and linked it with Pisa by means of a system of canals. Even later, however, warships went on taking to sea directly from the mouth of the river or sailing up-river without calling at the port (Rossetti, in print).

Inside the city, the infrastructure of the port consisted of shipyards, a wet-dock for galleys, a customs post and a merchant dock (Rossetti, in print) (Fig. 3).

TERZANA

The Terzana, dating back to the 13th century, was a wet-dock or harbour where the galleys could take shelter during the winter months; it was situated on the right bank of the river, beyond the bridge downriver, inside the trapezoidal area of about 4 hectares protected by the city walls built in the middle of the 12th century. No Mediaeval iconographical documentation seems to exist.

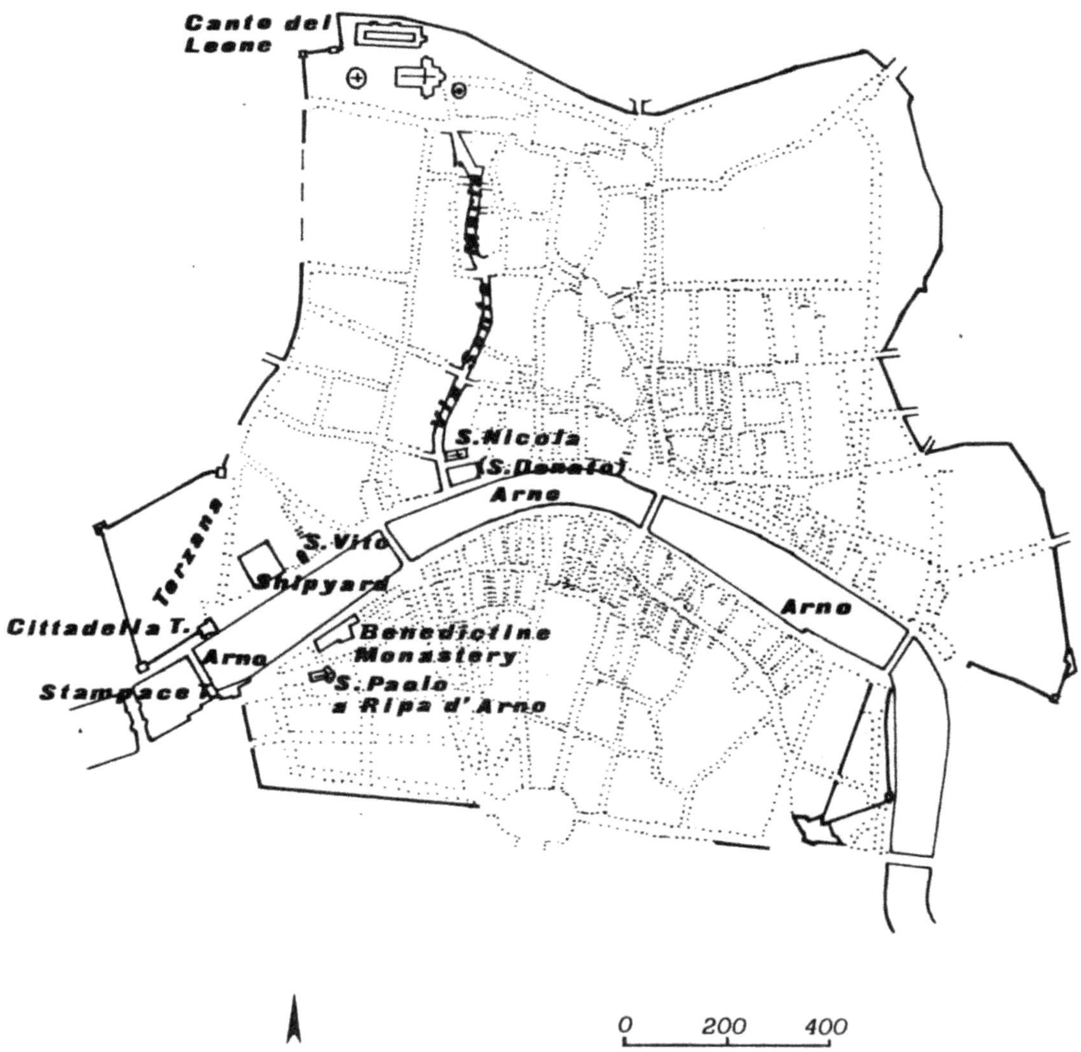

Figure 3: Pisa: Mediaeval harbour infrastructure.

The shipyards were right next to this area, between the old Degazia gate (beside the place where the Cittadella tower stands today) and the church of San Vito (the large sheds of the Medici shipyards, dating back to the 15th and 16th centuries, can still be seen in a small area next to San Vito). On the other bank of the river, the unbuilt-on area between the bridge down river and the Benedictine monastery which today is the Piazza of San Paolo a Ripa d'Arno, was also destined to become a shipyard, at least from the second half of the 13th century.

The outside walls of the Terzana are the only visible remains of the site today. Post-war restoration work led to the demolition of buildings and huts which, in the course of time, had been constructed, especially along the south wall (along the river); no structure above ground-level has survived inside the site; the Guelf tower no longer has its original shape, as a result of reconstruction and restoration work.

The south wall probably dates back to 1200, the year in which the walls were built around the Terzana, according to chronicle sources; curtain walls of the 14th century, and reinforcements added in the 16th century, are visible on the outside. The arcades facing inward, show that the sheds for the ships were arranged around a central dry-dock, to which the ships gained access via a canal, before being placed, stern first, inside the sheds. The 80 sheds for which a 14th century chronicler states there was room in the Terzana, do not apply to the original lay-out but to the period in which the site was at its largest (Rossetti, in print).

The north and west walls would seem to date back to not much earlier than the middle of the 14th century. The kind of materials used, the style and the construction methods are similar to those employed when the height of the city walls was increased for the second time (1250-1260).

The east wall (from the Canto del Leone, near the cathedral, as far as the Degazia gate) was the city wall built by Cocco Griffi in 1155; it was demolished in the middle of the 16th century since it no longer served any useful purpose.

Some uncertainty exists as to the date of the remains of a building with Gothic arches, located at the extreme western end of the Terzana; they have been variously attributed to the 14th and 15th centuries; today, the most probable date is considered to be the 14th century (Rossetti, in print).

The plan to build sheds in brickwork was already mentioned in the Statutes in 1286, a little less than 20 years after the construction of the north and west walls of the Terzana, which made the circle of walls complete. In such a brief space of time, the structures of the wet-dock capable of holding 80 large sheds, disappeared completely. It may be conjectured that these were wooden structures which were used again to build the fleet which was later to be defeated at Meloria (1284).

After the loss of Sardinia (1325), Pisa's maritime activities decreased continuously. The original structures, put up at a time of great expansion, must have been more than sufficient to meet the needs of the ships of Pisa, if they had not already disappeared. According to the political agreement with Genoa, Pisa was to use 10 galleys to guard the sea but it never employed more than three (Bonaini 1854-57, III, 745-746). While the building with Gothic arches, the only surviving structure of the Terzana as it was in Republican times, may be considered to date back to the time of Fazio di Donoratico (pre-1338), the Commune was so heavily committed in that century to the expenses for its army that the use of public money for structures of the wet-dock is unthinkable. In 1391, with a measure that aroused great hostility from the people of Pisa, Iacopo d'Appiano, the last lord of the Pisan Republic, transformed the wet-dock into a fortress, for his own safety, cutting off access to the Arno (Bellondi 1915, p. 16; Banti 1971). A few years later, occupation by the Florentines was to make the military use of the Terzana (which was meant to be only temporary) permanent.

After the Florentine occupation in 1406, the site was turned permanently into a military citadel. The other corner-stones of the defence system set up by Florence were the bridge downriver and the small new fort connected to the bridge, on the left of the Arno, called "Istà in pace" (of this, the Stampace tower and the ruins of the four towers of the citadel still remain (Rossetti, in print)). To ensure that such a defence system was safe and properly isolated, the areas of San Vito and San Paolo were destroyed after the Florentine occupation (Rossetti, in print).

Important iconographical documentation of the walled-off site that became the citadel and of the entire defence system created by the Florentines exists in a 15th century painting in the church of San Nicola in Pisa and in Vasari's fresco in the Salone dei Cinquecento in the Palazzo Vecchio depicting the victory of Florence at the siege of Pisa during the second war of independence fought from 1495 to 1509 by the Pisans after a century of subjection.

The anonymous author of the Pisan chronicle, dating back to the third decade of the 15th century, states that inside the citadel were to be found "...a church, a small loggia, as well as many houses and shops" (Supino 1904, p. 13).

Exceptional iconographical documentation is supplied by the map attributed to Giuliano di Sangallo in which it is possible to make out a galley, outside the walled area and near San Vito (further evidence of the fact that the shipyards were never inside the Terzana), and, inside, a church, houses, a loggia (so far the map tallies with the description given by the 15th century chronicler) and two sets of long shed on piles, with a canal leading to them which forks into two parts.

These two sets of shed may be probably considered to date to the years 1465-1468, the time attested to for their construction in Florentine records. They might have disappeared at the outbreak of the second Pisan war when the walls were again reinforced, or may have been left standing by the Pisans during the 15 years of war, in the hope of victory, and survived up to the time of Cosimo I when the building of new Medici dockyards began next to San Vito. Excavation and more accurate dating of the so-called Sangallo map will make it possible to put an exact date to their demolition.

CUSTOMS

The name "guardia sancti Viti" given to the shipyard area from the 12th century on, considerably earlier than when the Terzana "was built", is explained in a chapter of the Statutes of 1286 (Rossetti, in print). This is undoubtedly based on a by now well-established procedure concerning the customs office. Craft coming from or going to the sea had to be loaded and unloaded in the Arno, between the Degazia Gate and San Vito, and had to moor at San Vito (the "guardia sancti Viti") until customs formalities had been completed, without being approached by other craft. This was in order to make it impossible to avoid paying duty on a part of the goods (Bonaini 1854-57, I, Breve Pisani Communis, rubr. 59; II, rubr. 73). A decree by the Anziani of the city made it still compulsory in 1372 to pay duty on goods transported by cart and by boat from Porto Pisano to Pisa and vice versa (this means along the canals and the Porto Pisano road, as will be discussed later) (Arch. Stato Pisa, Commune, Div. A, Reg. 197, pp. 127-8, 1372 October 27). This does not necessarily mean, however, that bigger craft than the small ships and rafts which were towed up the canals, loaded with light bulky goods such as wool, might not have gone up the Arno from its mouth. This would have been under the offshore surveillance of the ships involved in the guardia maris, supplementing the internal linkage system that had become unable to cope with the increase in the volume of traffic from the 13th century on.

The Customs officers escorted the craft that proceeded under the downriver bridge (a draw-bridge) up to the stretch of water at San Vito where they went through Customs controls and then paid duty. From here, perhaps on boats operating a ferry service, the goods were allowed to go on their way to the merchant harbour, passing under the New Bridge.

WET-DOCK

The inner wet-dock was located in the area between the church of San Nicola and that of San Donato (which was subsequently to form part of the east side of the then grand ducal palace). Here, the goods were put on sale at the market held in Piazza San Donato, or they were sorted out in the warehouse, a magna domus pro communi utilitate habens caput in Arno et iuxta viam Sancte Marie, the building of which began in 1160, no doubt as a warehouse, and to house the trade offices of the city port (Pertz 1963, s.a. 1160, p. 246). From here, the goods were sent overland to their place of destination, or were loaded on to craft plying up and down the river to and from Florence, as well as to the villages scattered along the lower Arno valley or along the Emilia highway, which was the most densely populated route of the Pisa district and surrounding area and the one most used for movement towards the city.

PORTO PISANO

Praising Pisa and its port immediately after the Florentine occupation of the city (October, 1406), Goro Dati describes Pisa's geographical position, situated five miles from the sea and 12 from its port, with the Arno running through it used by boats loaded with goods, with the citadel at its farthest western point, on the right bank of the river, and its sea port "excellent and large, enclosed by towers with a chain between them, near Livorno". These details give the cardinal points of a system of installations and coordinated infrastructure, geared to an economic system and a complex market, but at a time when its maritime importance was waning, even though its access to the sea and its port were not yet a thing of the past towards the end of the free respublica pisana.

Porto Pisano was a vital lifeline for the city, with the extensive infrastructure described above (dockyards, wet-docks for galleys and for merchant craft) and all the relevant activities (commercial, industrial and service). An efficient network of roads and canals maintained daily contact between the port and the maritime infrastructure in the city and the organization of labour was such that they were dependent upon one another.

The author of the 13th century Compasso de navigare describes Porto Pisano as being a "porto di catena" which is protected but also endangered by shallows. On one of these shallows, 10 miles southwest, was the Meloria tower, and, to the east, the Fanale tower. Craft entering the port needed to be especially cautious. The ship had to be well in line with the two towers guarding the entrance, in order to pass safely and avoid the shallows, with Monte Nero to the east.

Geomorphological survey makes it possible to establish that the access route the author of the Compasso advises navigators to take, runs north of the Gronda dei Lupi (Mazzanti, Pasquinucci, Salghetti Drioli 1984). The towers of the port, which no longer exist, were situated along this line.

The author of the Compasso also mentions Bocca d'Arno and the fact that narrow craft could enter. This involves something which was a perpetual problem for navigation on the river, the need to keep the mouth constantly free of debris which made the waters shallower, in order to allow navigation by narrow ships with limited draft, such as the galleys that wintered in the Terzana.

As has already been noted, Porto Pisano was equipped to meet the needs of rapidly expanding trade in the middle of the 12th century, in a decade of fundamental importance for the history of the city-state of Pisa.

Fernando Maragone's Annales (Pertz 1963) and the Breves consulum for 1162 and 1164 (Bonaini 1854-57) give us a painstaking account of what was done. While the city walls were being built (1155-1161), the construction of the Meloria tower was begun, opposite Porto Pisano (1156). In 1158, five galleys were fitted out to guard the sea and construction work began on the two towers guarding the main area of the harbour. A fountain pro utilitate marinariorium was got ready at Santo Stefano ai Lupi (Pertz 1963, s.a. 1158, p. 244); and in 1162, while the magna domus pro communi utilitate (mentioned earlier) was being built in the city, a large establishment pro utilitate marinariorium was set up at the port and a well was dug nearby (Pertz 1963, s.a. 1162, p. 247).

1160 saw the excavation of the first 50 pertiche of the Vettola canal. This reached San Piero a Grado apparently following the course of a former branch of the Arno. In 1162 work was about to start on the first 50 pertiche of the Caligi di Stagno (from Stagno to the Arno, about 8.70 m wide at the top and 7.30 m wide at the bottom, to a depth of roughly 1 m; Bonaini 1854-57, I Breve consulum, p. 37); 50 pertiche of the San Piero canal were also dug (from San Piero a Grado to Stagno, a continuation of the Vettola canal (Bonaini 1854-57, I, Breve consulum p. 39).

Moreover, the towers were fortified and the port was closed off with a great chain between them (Pertz 1963, s.a. 1162, p. 247). In the meantime, a running war had broken out with Genoa). All the work undertaken had been completed by 1164.

Ten years later, a warehouse for goods arriving and being dispatched was added to the harbour installations (Pertz 1963, s.a. 1162, p. 247; Rossetti, in print). The archbishop of Pisa also helped in equipping the port with its essential infrastructure by founding the hospital at Stagno (1164) to replace the old hospital at Uscione, and by undertaking the construction of the bridge linking the harbour installations, the Porto Pisano road and the city. Another purpose of the new structure was to ensure that the port had the necessary capacity which it never actually attained (Rossetti, in print).

From this, we can see that while the ships were built and fitted out in Pisa, they sailed empty as far as Porto Pisano. This made it easier for them to leave by the mouth of the river (in 1162 the Genoese had taken three empty ships by surprise at Bocca d'Arno and set fire to them). Men and goods were put on board at Porto Pisano after having been transported there by barges and small craft along the canals, or by cart along the Porto Pisano road.

It is hard to believe that two-way traffic was possible on the canals, given their width. The Stagno waterway, linked to the Arno to the east of Pisa (between Putignano and Oratoio), seems likely to have been used to transport goods away from the city; the San Piero-Vettola canal, connected to the Arno west of the city, was used for incoming goods. Indeed, payment of duty was made at the bridge down river.

At the end of the 13th century, the port was guarded by four towers: the Soarsa and San Vito towers, linked by the chain, stood at the entrance of the access canal, the Frasca tower was further to the west, while the Magnale tower lay farther back at the centre. To the east of the Livorno rock stood the Fanale tower and, five miles out to sea, that of Meloria.

WAREHOUSE

The warehouse, complete with a tower and an iron gate, was to have been situated at the extreme left of the port, according to the inscription on the marble tablet the Genoese Nicolao de Guilielmo had had carved in 1290 to celebrate the feat of Corrado Doria, the victorious admiral. In an imaginary line-up, but one with a perspective likely for someone approaching the port by the access canal, are engraved, from left to right, the Fanale tower (with a large torch burning in an iron cage), the two Soarsa and San Vito towers, a portico (attested to in the year 1268 but which must have existed for some time at that date), the warehouse and, some way off, the Magnale tower (Rossetti, in print).

In 1290, only 6 years after the Meloria battle, the Genoese seriously damaged Porto Pisano. The troops of the Guelf League occupied the area and Livorno, blocking the entrance to the largest canal by sinking a galley in it, de matonibus plenam et muratam, transported there from Genoa for the purpose. They irreparably lowered the depth of the water by tipping the rubble from the dismantled towers into the canals (Rossetti, in print). The Fanale tower, which could not be captured, was made to collapse into the sea.
It had already been rebuilt by 1305, Petrarch describing it with admiration in his Itinerarium Syriacum; at the beginning of the 15th century, Goro Dati called it the world's most beautiful tower. It was destroyed during the Second World War and later reconstructed in its original form (Rossetti, in print).

Time brought no great change to the structure of the port compared to what it was in the middle of the 12th century, whereas the problems facing Porto Pisano increased, with the threat of being silted up becoming greater and greater until the time of the disastrous blockage caused by the Genoese in 1290. At the same time, the mouth of the Arno had to be dredged continuously because the river-bed rose constantly.

What did change, however, during the 13th century was the whole system of ports which was adapted to the demands of an increasing maritime activity that Pisa and Porto Pisano could not meet, since their capacity was objectively limited. In the second half of the century, Livorno, Piombino and then Vada played a supporting role as landing places, for recruiting men, loading provisions and repairing hulls. Thanks to its mooring space (enlarged and safer after the construction of the township's defences in the first half of the 14th century, and owing to improvement of the docks), Livorno gradually found itself taking over from Porto Pisano in every respect and excluding it from the system of ports.

Pisa retained its ship-building role for one very good reason. From as early as the beginning of the 12th century, the supply of wood for the shipyards from Garfagnana, Mugello and Versilia passed along the inland waterways, the Serchio, the Arno and their tributaries (Calisse 1904, vv. 98 ss., pp. 10-11).
Above all, Pisa was and was to remain, the sole commercial harbour and the sole market for the sorting of goods, a monopoly which was strenuously defended (Bonaini 1854-57, II, Breve Pisani Communis, I, rubr. 122; III, rubr. 111). The creation of new ports endangered public income and the system of taxation.

Porto Pisano continued to operate throughout the 14th century, without ever recovering its former efficiency while Livorno's size as a port and as a community grew. Pope Gregory XI's journey with his retinue from Avignon to Rome, in 1376, was typical in this respect (Rossetti, in print). The first part of the convoy moored at Livorno, where 7 cardinals stayed; the second part, with the Pope, stopped off at Porto Pisano, but not to stay there and proceed to Pisa: the Pope's galley took to sea again to moor at Livorno. Pietro Gambacorti, the lord of Pisa, was there to meet him and escort him to the parish church (of Livorno).

This shows Livorno's capacity as a port and as a community such as to provide suitable accommodation for high-ranking people. Porto Pisano is described by a chronicler of the time, Ranieri Sardo, as being virtually part of Livorno. This is tantamount to saying that the port's main centre was commonly held to be Livorno, not Pisa. No mention is made of the latter, which now seems to be excluded from the port system.

The Florentine occupation of 1406 and a long period of inactivity marked the decline of Porto Pisano. Pisa, described by Goro Dati with amazement and admiration immediately after the conquest of 1406, was beyond recognition, indeed only one generation later.
No other city, except Venice (which was not vanquished) owed the survival of its social and economic structure, as Pisa did, to such a finely balanced set of circumstances, obtained by means of the very great and continuous financial commitment made by the Town Council in the form of public works.

For at least 300 years, from the 12th to the 14th centuries, the regulation and maintenance of the waterways in the city, in the surrounding area and in the harbour area was by far the biggest item in the budget of the Pisan Republic. The most assiduous attention was paid to maintaining and progressively improving the efficiency of an intricate system of canal works and installations for the reclamation of marsh-land, facilitating connections, harnessing the fast-flowing waters of numerous tributaries of the Arno and the Serchio as well as dredging the mouth of the former and the canals of the port and repairing the bridges and defences.

This facet of the problems must be properly emphasized if we wish to understand the "mystery", hitherto never completely cleared up, of the sudden decline and total ruin of Pisa under Florentine domination. It was certainly not the only city in Italy to be dominated but it was certainly unique in coming to such a quick and tragic end.

What changed the fall of Pisa into irreversible disaster was, in my opinion, not so much the departure en masse of the upper classes and the unbearable pressure of taxation (Rossetti, in print) but rather the neglect of public works by the Florentines over a period of more than sixty years. When, as inevitably happened, taxation and social pressure eased, the city was not longer capable of producing, nor the surrounding area able to find a remedy for its desolation.

Indeed, for at least 60 years after Pisa capitulated, Florence's sole preoccupation, as far as the city was concerned, seems to have been to fortify it in order to be sure of getting a firm hold on it (Silva 1909, 133-183, 285-323; 529-570, app. n. 3; Rossetti, in print). Florentine policy was to run as little risk as possible and to exert extreme harshness in the treatment of the conquered, with irreversible effects on Pisa's urban fabric, society and economy. Florence destroyed whole areas of the city (the districts of San Vito and San Paolo mentioned earlier), persuaded the upper classes to choose exile for good to avoid the internal exile imposed on them by the invaders, as well as seeing their property and wealth confiscated. Those who stayed on were forced to stop all financial activity, even if they managed to avoid internal exile. This, together with confiscation, destruction by the soldiers and exceedingly onerous taxation, dried up Pisa's resources and finally made recovery impossible (Silva 1909, p. 168, app. no. 3; Fanucci, Lovitch, Luzzati, 1986).

In 1421, the Florentines were able to purchase Porto Pisano and Livorno from the French, who were holding them at that time.
The Signoria took steps (Grunzweig 1930) to revitalize maritime activity, following the example of the Venetians and their "galley auction". Ship-building was resumed at once, after an interval of at least 28 years (Bongi 1892, III, p. 304; Mallet 1967) and skilled workers were brought in from Venice and Genoa. The space available for building (after the events and changes described above) was in the area next to San Vito, which had become free as a result of the destruction of the houses there, and which was used for the Medici shipyards set up in the 16th century (95-97) (See the Memoriale by Portoveneri; and the map of Giuliano da Sangallo, mentioned earlier).

The state of the Arno turned out to have deteriorated considerably: the journey down to the mouth described by Luca di Maso Degli Albizzi (1428) lasted three days, the ships ran aground and were towed by men, using capstans. After the cargo and crew had been put on board at Port Pisano a single day's sailing was enough for them to reach Monaco (Mallet 1967).

Wide scale restructuring work on Porto Pisano began in 1439. The towers were rebuilt and fortified, a "clausura palorum...volgarmente detta la palicciata" was created, a reservoir was installed and the canal leading into the port was dredged. Repair works and restructuring were also begun on the "little harbours" of Livorno, and the access canal was made viable so that even large galleys could enter. These steps, taken during the rule of Cosimo the Elder, give some idea of the policy of increasing maritime activity and of constantly adapting the structure of the shipyards and port for this purpose, that was also adopted by Piero and Lorenzo dei Medici, during the 15th century.

In Pisa, when the "old" Cittadella no longer served any military purpose, it was used partly as a wet-dock and safe harbour for galleys (1468) (Gaye 1840, I, app. II, pp. 569-570). The shipyard remained near San Vito.

It was Lorenzo the Magnificent who realized that the reason for the very serious deterioration of the situation in and around Pisa was the failure to keep the ditches, roads, bridges and sewers in good working order and he took steps to remedy the matter and reestablish a balanced environment (Fiaschi 1938). But his plan could not be put into practice because of the fall of the Medici and the outbreak of the second Pisan war.

Restoration work started again when the principality was founded, but this is another story, no longer centred on Pisa and Porto Pisano but rather on Florence and Livorno. Pisa had by this time been excluded from the whole system and the port was Livorno.

Two thirds of the area of the old port, from the Gronda dei Lupi to the mouth of the Calambrone, are now taken up by the Livorno shipyard. Compared to the 13th century the coastline is at present almost three kilometres further forward. The Florentine Marzocco tower (1458-1474), all that remains of the structure of the Mediaeval port and which was formerly more than a mile out to sea, is now surrounded by a concrete parking area for containers, 500 metres long, by the coast road, and by the port's oil refineries. The very name of Porto Pisano has disappeared, but life goes on.

REFERENCES

Bellondi, E. 1915. Cronica volgare di anonimo fiorentino dall'anno 1385 al 1405. Bologna (Rerum Italicarum Scriptores 27).

Banti, L. 1943. Pisae. Memoria Pontifica Acc. Rom. Archeologia, Ser. 3, 6.

Banti, O. 1971. Jacopo d'Appiano. Economia, Società e Politica del Comune di Pisa al suo Tramonto (1392-1399). Livorno, S.E.I.T.

Bonaini, F. 1854-57. Statuti Inediti della Città di Pisa dal XII al XIX Secolo, I-III. Firenze, Viesseux.

Bongi, S. 1892. Le Croniche di Giovanni Sercambi Lucchese, Pubblicate sui Manoscritti Originali, I-III. Roma (Fonti per la Storia d'Italia).

Calisse, C. 1904. Liber Maiolichinus de Gestis Pisanorum Illustribus. Roma (Fonti per la Storia d'Italia, 29).

Ciampoltrini, G. 1981. Il monumento dell'augustale Constans a Lucca. Prospectiva 25: 37-42.

Ciampoltrini, G., Cianferoni, G.C., Romualdi, A. 1982-83. La Raccolta Archeologica e Numismatica Enrico Chiellini, il Museo Archeologico di Livorno ed i Materiali dal Portus Pisanus. Rassegna di Archeologia 3: 183-241.

Fanucci, M., Lovitch, L. and Luzzati, M. 1986. L'Estimo di Pisa nell' Anno del Concilio (1409). Pisa, Pacini.

Fiaschi, R. 1938. Le Magistrature Pisane delle Acque. Pisa, Nistri e Lischi.

Finzi Contini, G. in print. Indirizzi per indagini geofisiche affiancabili ad elaborati da telerilevamento nell'area di Portus Pisanus. In: La Gestione del Suolo e delle Acque (Atti Conv. Naz. I.R.O.E. Firenze 1986).

Gaye, G. 1840. Carteggio Inedito d'Artisti dei Secoli XIX-XVI. Firenze, Molini.

Grunzweig, A. 1930. Les fonds du Consulat de la Mer aux Archives de l'Etat à Florence. Bulletin Inst. Historique Belge, Rome.

Lugang, R. 1926. Note sur l'itinéraire maritime de Rome à Arles. MEFRA 43: 124-139.

Mallet, M.E. 1967. The Florentine Galleys in the Fifteenth Century (with the Diary of Luca di Maso degli Albizzi Captain of the Galleys 1429-1430). Oxford, Clarendon Press.

Mazzanti, R. and Pasquinucci, M. 1983a. L'evoluzione del litorale lunense-pisano fino alla metà del XIX secolo. Bollett. Soc. Geografica Ital. 12: 605-628.

Mazzanti, R. and Pasquinucci, M. 1983b. The evolution of the Luni-Pisa coastline (II-I cent. B.C.E. - second half of the XIX cent.): archaeological, documentary and geomorphological evidence. In: Coastal Problems of the Mediterranean Sea (Intern, Geogr. Union Commission on the Coastal Environment) E.C.F. Bird and P. Fabbri (eds.). Bologna.

Mazzanti, R. and Pasquinucci, M. La costa tirrenica da Luni a Portus Cosanus. In: Déplacement des Lignes de Rivage en Méditerranée d'après les Données de l'Archéologie (Actes Colloque Internat. C.N.R.S., Aix-en-Provence, Marseille, 1985). In print.

Mazzanti, R., Pasquinucci, M. and Salghetti Drioli U. 1984. Il sistema secche della Meloria-Porto Pisano: geomorfologia e biologia marina in relazione ai reperti archeologici. In: 1284: L'Anno della Meloria. Pisa, E.T.S.

Pasquinucci, M. 1986. Rinvenimenti nell'area di Coltano. In: Terre e Paduli: Reperti Documenti Immagini per la Storia di Coltano. R. Mazzanti, R. Grifoni Cremonesi, M. Pasquinucci (Eds.). Pisa-Pontedera, Bandecchi e Vivaldi.

Pertz, K. 1963. B. Maragonis Annales Pisani 1004-1175 (M.G.H. SS. 19).

Romualdi, A. 1982. Livorno. Bollettino Arte, Suppl. Archeologia subacquea, 4:55-56.

Rossetti, G. in print. Pisa: assetto urbano e infrastruttura portuale. In: Città Portuali del Mediterraneo (Congresso Genova 1985).

Schmiedt, G. 1964. Contributo della Fotointerpretazione alla Ricostruzione della Situazione Geografica-Topografica dei Porti Antichi in Italia. Firenze, Ist. Geografico Militare.

Schmiedt, G. 1978. I porti italiani nell'Alto Medioevo. In: XXV Settimana Studio Centro Ital. Alto Medioevo. Spoleto.

Silva, P. 1909. Pisa sotto Firenze dal 1406 al 1433. Studi Storici 18: 133-183.

Supino, I.B. 1904. Arte Pisana. Firenze, Barbera.

Targioni Tozzetti, G. 1768. Relazioni d'alcuni Viaggi Fatti in Diverse Parti della Toscana. II. Firenze, Stamperia Imperiale.

Toscanelli, N. 1933-34. Pisa nell'Antichità, dalle Età Preistoriche alla Caduta dell'Impero Romano, I-III. Pisa, Nistri-Lischi.

PAOLO ANTONIO PIRAZZOLI

Sea-Level Changes and Crustal Movements in the Hellenic Arc (Greece). The Contribution of Archaeological and Historical Data

INTRODUCTION

Varied evidence has been found recently to show that a major tectonic event, called the Early Byzantine Tectonic Paroxysm (EBTP), unprecedented in size for at least the last 100,000 years, took place in the eastern Mediterranean about 1530 yr BP (Pirazzoli, 1986). An almost uniform uplift of 0.6-0.8 m occurred along the Levantine coasts and near Alanya, Turkey. The most drastic tectonic movements though are reported from western Crete and from Antikythira Island, where a series of ten rapid subsidences of a huge block of lithosphere without noticeable tilting between 4000 and 1700 yr BP, was followed about 1530 yr BP by a sudden tremendous uplift of about 10 m, and a northeastward tilting (Fig. 1).

In this paper, a review is given of the available archaeological and historical data which indicate relative sea-level changes in the Hellenic arc. In western Crete and Antikythira most of the data, as expected, are in full agreement with the known tectonic pattern. In other parts of the Hellenic arc, where some shorelines had previously been radio-carbon-dated, archaeological data are the main evidence of vertical displacements which, in some cases, may be related to the EBTP or to its preparatory phase.

PHYSICAL SETTING OF THE HELLENIC ARC

The Hellenic or Aegean arc is situated in the southernmost part of Greece at the boundary between the Aegean and the Mediterranean Seas. It extends between the Ionian Islands and the Anatolian coast, and includes the western Peleponnesus, Kythira, Antikythira, Crete, Kassos, Karpathos and Rhodes (Fig. 2). The latter islands, spread along a 700 km curve, approximately follow an arc of a circle of about 400 km in radius centred on the Aegean Sea.

The Hellenic arc shows all the morphological characteristics of an insular arc; it is associated with an inner volcanic arc (Aegina, Milos, Santorin, Nisiros), about 250 km in radius, a marginal basin (the Sea of Crete) and, on the outer side, a system of trenches (the Hellenic Trenches).

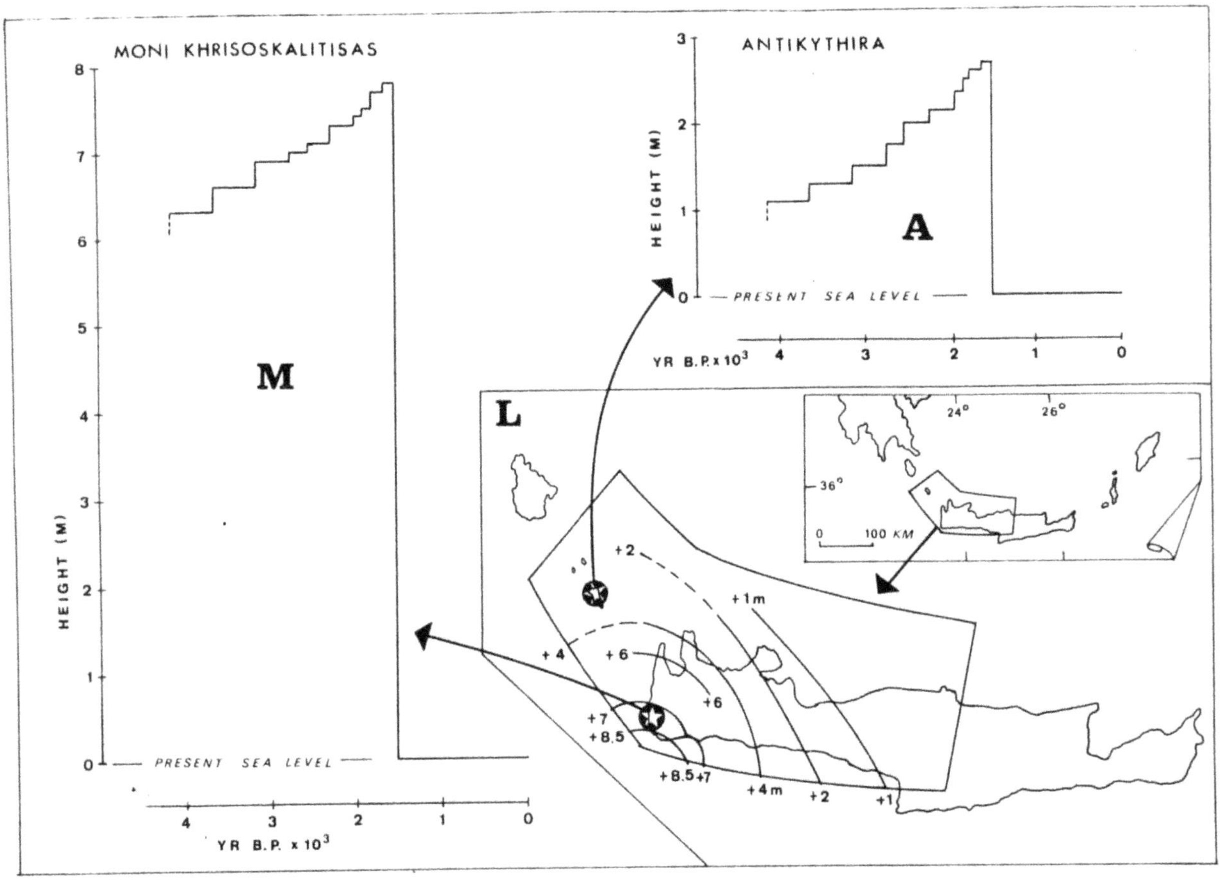

Figure 1: L= approximate boundaries of a lithosphere block about 200 km long which was uplifted and tilted in the Hellenic arc about 1530 yr BP., with elevation contours of the upheaval in metres. M= relative sea-level variations at Moni Khrisoskalitisas A= relative sea-level variations in Antikythira Island.

The straits between the Sea of Crete and the Mediterranean are relatively shallow; the 500 metre isobath shows only one narrow passage northeast of Karpathos and the 900 metre isobath, a continuous arc from the Peloponnesus to Anatolia. The relief of the islands is often significant: 1200 m in Karpathos and in Rhodes, and three massifs reaching between 2000 and 2500 m in Crete.

A clear morphological dissymmetry exists between the outer and the inner parts of the Hellenic arc, a shape that has been called by Angelier (1979) "toit en usine" (single pitch roof), the difference in elevation between the top of the relief and the sea floor being much larger on the trenches side than on that of the Sea of Crete.

The climate of the area is typically Mediterranean, with a dry and warm season (from May to October), a cool and moist season (from December to February) and transition conditions in the other months. Temperatures are usually mild, all the southern part of the Hellenic arc being situated south of the 19 isotherm. Winds most frequently blow from SSW between November and March and from NNW from April to October. The summer winds, called Etesian, are fairly regular and constant from the middle of July to the end of October. The tidal range is usually less than 0.2 to 0.3 m.

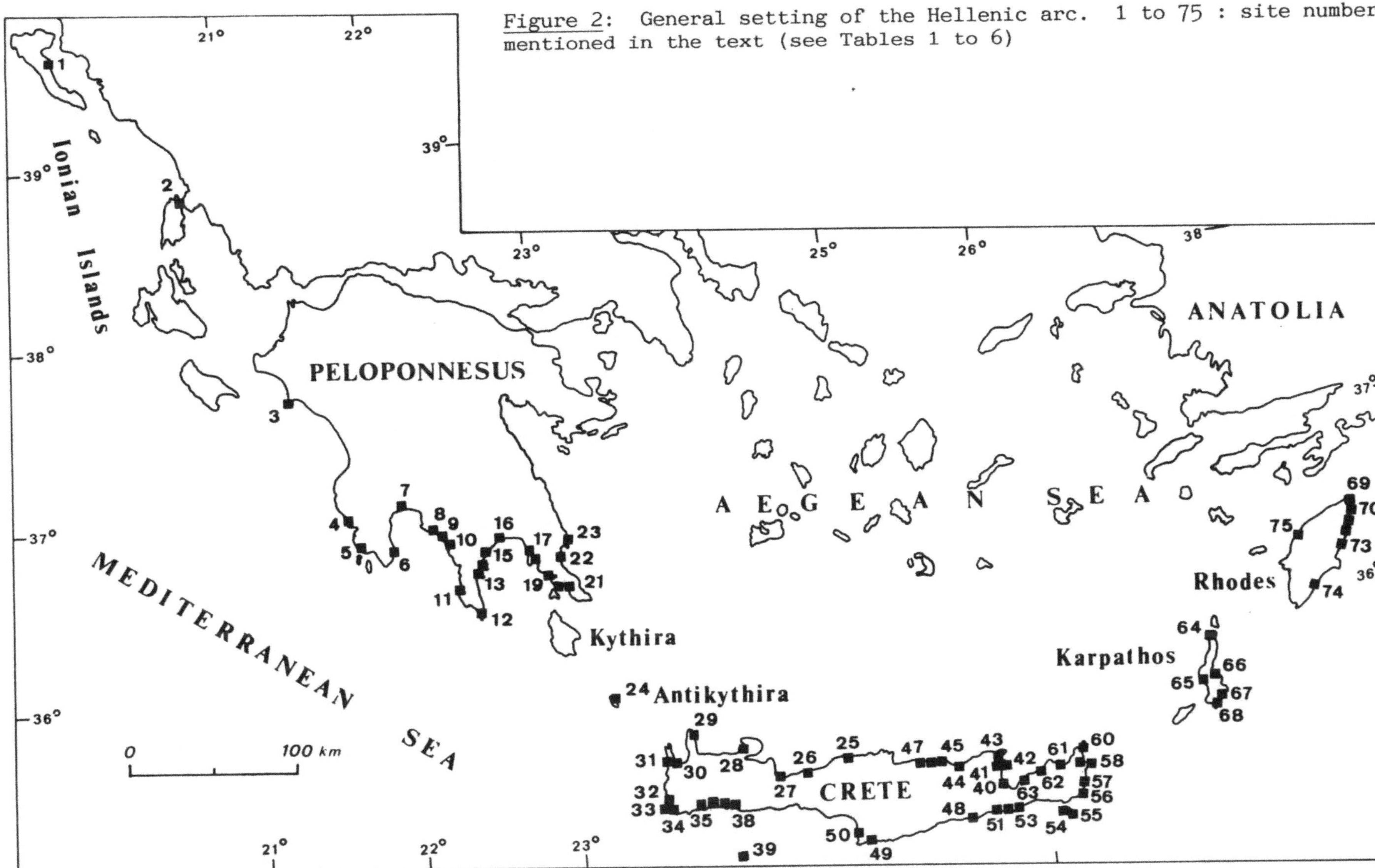

Figure 2: General setting of the Hellenic arc. 1 to 75 : site numbers mentioned in the text (see Tables 1 to 6)

PREVIOUS STUDIES AND INTERPRETATIONS

Previous reports on changes in sea-level during historical times date back to ancient times in Crete, where the Venetian, Onorio Belli, observed the submerged remains of the towns of Olous and Minoa in 1586 (Spratt, 1865, p. 138-141).

Credit for the discovery of emerged shorelines in western Crete seems to be due to Lieutenant Mansell (Leycester, 1852) and Captain Spratt (1865) who went on a British Navy reconnaissance of the coasts of the island. Spratt not only identified and measured the altitude of marks of ancient shorelines in many localities, but was also the first to understand that the upheaval probably occurred in historical times. According to him, the upheaval was not the result of a single movement, but of a series of jolts.

Spratt's results influenced Raulin (1869), who proposed an interpretation (which was later wrongly attributed to Spratt), according to which "the whole of Crete" would have made a "tilting motion which lifted up the western third and lowered the eastern half" of the island. This interpretation, perhaps because it was received favourably by supporters of mobilistic theories such as Issel (1883), was criticized by defenders of the dominant stability theories. Thus, when Suess (1888, p. 741) claims that "the Mediterranean shows no evidence of slow uplift or subsidence movements of the lithosphere dating from historical times", Cayeux (1907) echoes this with "western Crete shows no evidence of shoreline changes during historical times."

This statement by Cayeux is a remarkable example of scientific blindfolding, justifying the defence of pre-established theories. Thus, after having visited the submerged remains of Spinalonga reported by Issel (1883, p.279), Cayeux still affirms (1907, p.114) "not having been able to state any certain mark of recent uplift or subsidence."

Negris (1904) on the other hand, although agreeing with the conclusions of Suess and Cayeux on the "stability" of western Crete, nevertheless ascribes the submergence of eastern Crete to a gradual sea-level rise. According to Negris, "after the submergence of Hellenic buildings, that of Roman buildings, then of Christian churches and lastly of buildings of the Venetian dominion took place."

These statements of supporters of stability theories seem to have had a sterilising effect on research in this field, after Cayeux (1907), more than half a century was to elapse before new contributions were made.

It is a matter of fact that it was only in 1965 that Hafemann published the first radiocarbon datings confirming that the uplift of western Crete had occurred in historical times. Later, especially during the 1970's, several new measurements and datings were carried out by B. Keraudren and J. Angelier (see Laborel et al., 1979, Table 2), Flemming *et al.* (1973a,1973b), Flemming (1978) and Kelletat (1979). Flemming (1978), in particular, developed a geodynamic model of the northeast Mediterranean based on the assumption of unidirectional step-by-step crustal movements during historical times (of uplift in Antikythira, western Crete and Rhodes; of subsidence in Karpathos), thereby challenging the stability theories.

At about the same time a new systematic survey, corroborated by a multi-disciplinary approach, enabled Pirazzoli (1979a), Laborel *et al.* (1979), Thommeret *et al.* (1981) and Pirazzoli *et al.* (1981, 1982a) to demonstrate that the uplift movement in western Crete and Antikythira did not occur gradually by jolts as was believed by Spratt and Flemming, but by a sudden single jerk around 1530 yr. BP, after a series of small subsidence movements during the preceding two or three thousand years (Fig. 1).

In central and eastern Crete, although subsidence phenomena have been predominant, crustal movements appear to be much more irregular than the rigid tilting of the island assumed by Raulin (Flemming, 1978; Flemming and Pirazzoli, 1981; Pirazzoli *et al.*, 1982a).

During the last two decades, the very rapid progress of neotectonic and geophysical research in the Aegean region has been of great help in understanding the possible causes of recent shoreline changes, in the framework of the tectonic and isostatic processes which are occurring in this very unstable Mediterranean area.

ARCHAEOLOGICAL INDICATORS OF SEA-LEVEL CHANGE

Submerged remains have been identified since antiquity as marks of relative movements between the land and the sea. They were, until a few decades ago, the only manner in which to estimate and date these movements. In this context the Aegean region is especially well situated, with its ancient civilization and the important seismic, volcanic and tectonic movements affecting it. Overviews were attempted at the end of the last century, in order to interpret crustal movements from the data available so that it is not surprising that the regional work of Spratt (1865) in Crete precedes the global approaches by Issel (1883) and Suess (1888).

In the Mediterranean, research studies on sea-level changes were based for a long time on compilation rather than on systematic field checking. It is only with Flemming (1969) that a systematic visit to many sites was made in the western Mediterranean. For the first time, many sea-level indicators were investigated *in situ* and compared with one another by the same observer. Later, this pioneer work was extended to the Aegean coasts, Cyprus, southern Anatolia and the Hellenic arc (Flemming *et al.*, 1973a, 1973b; Flemming, 1978; Flemming and Pirazzoli, 1981).

Meanwhile, detailed reviews were devoted to specific topics; ancient harbours (Blackman, 1973, 1982a, 1982b), fish tanks in Italy (Schmiedt, 1972, 1981) and in the Mediterranean (Pirazzoli, 1979b), and all the archaeological indicators of sea-level change (Flemming, 1979).

According to a classification outlined in earlier papers (Pirazzoli, 1976; Flemming, 1979) three main categories of archaeological remains can be distinguished;

1. those having no clear relation with sea-level;
2. those which must have been located near a shoreline and, according to their use, must have remained emerged or submerged;
3. lastly, those belonging to structures built partly underwater and depending for their use on tidal fluctuations and marine conditions.

The remains belonging to the first category are the most frequent. They may be used as sea-level indicators only if they have been covered by sea-water since their construction. Remains of this type have been observed in several localities of the Peloponnesus, in central and eastern Crete, in Karpathos, and on the southwest coast of Rhodes. They indicate a rise in the relative sea-level equal to or greater than the present submergence, although the change in elevation cannot be estimated precisely.

Partly submerged coastal quarries, such as those in Karpathos and Rhodes, belong to the second category of remains. The blocks cut from quarries were most frequently transported to their place of use by boat much more easily than by land. Owing to this, layers good for quarrying were worked as much as possible near the loading area. From what is known of quarrying techniques in antiquity (Trousset and Guery, 1981), the work could only be carried out correctly out of the water and the deepening of coastal quarries had to stop just above sea-level. This means that for quarries which are presently submerged, the change in sea-level can be estimated with accuracy.

However, caution may be necessary in interpretation. In some sites, a few metres behind the shoreline, deep quarrying appears to have continued to a lower level, leaving a slightly higher seaward barrier of rock, probably intended to make landing by boat easier and to protect the quarry from wave projections. Thus, if the rock was not too porous, sea water could not penetrate inside the quarry and rock cutting could be carried out slightly below sea-level. Later, if the barrier broke, sea water would flood the floor of the quarry and submergence would appear greater.

When the floor of a coastal quarry is almost horizontal and well above the present sea-level, this does not necessarily indicate a sea-level drop, as it may have been considered preferable to quarry certain exposed, almost horizontal layers of the rock.

The second category of indicators also includes remains which are at present above the surface, but which had been constructed entirely underwater, or remains constructed partly underwater, but with the upper part now missing. These vestiges (an example is the breakwater in the Roman harbour at Kisamos, Crete) make it possible to estimate the minimum drop in the relative sea-level.

Indications, unquestionably more precise, are provided by the third category of remains, which enable margins of uncertainty in relative sea-level change to be estimated. Remains belonging to this category include slipways (e.g. in Antikythira, Rethimnon, Karpathos), certain harbour constructions (e.g. in Falasarna) and especially fish tanks. When their most characteristic parts are still recognizable, Roman fish tanks provide an excellent possibility of accuracy in the estimation of the age and extent of a sea-level variation.

Varro (116-27 B.C.E.) states in his De re rustica that fish tanks were constructed more often near villas (3.2.3.) and that the fashion of marine fish tanks started with his generation (3.2.10). At that time freshwater fish ponds were little esteemed and left to the plebs, whereas sea water piscinae were prized and considered as a real luxury. They did not bring any profit and even helped to empty the purses of their owners rather than to fill them(17.2.2.); costs of construction, stocking and maintenance were in fact very high. The most prized fish tanks consisted of several basins, making the separation of fish species possible (17.2.4.). The costs were so high that no cook would have dared prepare these fish. That is why Quintus Hortensius (114-50 B.C.E.), who spent much money on feeding fish in his piscina, used to purchase fish for his table at Pozzuoli market (17.2.5.).

Columella (1st c. A.D.) in his De re rustica gives several details on fish tank construction. Sea water must be renewed as much as possible by waves (17.1). The basins may be cut into the rock or constructed along the coast with walls made of broken sherds (opere signino). Anfractuosities may be arranged on the floor, some of them straight and regular for scale-fish, others narrow and meandering to shelter Murenae (17.2). When the coastal shape allows it, lateral openings make water circulation easier. A reminder of the need to allow for water to be replenished is often given by Columella. In order to protect fish from the sun's heat, recesses are to be created on the walls of the basin (17.6). At the entrances of supply channels, bronze grids with a fine enough mesh to prevent fish from swimming away, have to be provided (17.7). The walls of a basin must emerge at all times, even when the sea is rough. Outer walls slightly higher than the inner walls have to be constructed around the fish tank, in order to protect it from storm effects (17.10). Lastly, in some places the outer protection walls must be cut by narrow passages, in order to enable sea water to come in, though stopping the waves (17.11).

The fashion for fish tanks must have seen a rapid boom through-out the Mediterannean, since Pliny (A.D. 23-79) mentions them in places as far distant as Narbonne and Phoenicia. Many fish tanks were constructed on the coasts of the Tyrrhenian Sea near Rome, where remains of very complex and vast fish-tank installations have been reported (Schmiedt, 1972).

The short-lived fashion for fish tanks (approx. 1st century B.C.E. - 1st century C.E.) enables the related sea-level to be dated with accuracy. In the Hellenic arc remains of fish tanks have been found mostly in Crete, at Falasarna (see below), Khersonisos, Mokhlos, Sitia, Kato Zakros and Ierapetra (Pirazzoli, 1979b; Flemming and Pirazzoli, 1981). Some of their construction elements give very precise indications of the position of the former sea-level, for example, the channels open to the sea had to supply water at least at high tide; the height of the basic walls had to prevent fish from escaping, although allowing wave splash to come in; vertical grooves cut into the rock to receive sliding slabs or gratings must have reached a level above high tide.

DATA

The Northwest Part of the Hellenic Arc (Fig. 2, loc. 1 to 23)

Northwest of Crete, behind the Ionian trenches which stretch towards Apulia in an almost straight line, the Hellenic arc partly loses its island character with the intrusion of the Peloponnesus between Kythira and the Ionian Islands.

Few archaeological remains are mentioned in relation to sea level in the Ionian Islands and in the northwest Peloponnesus. In the southern part of the Peloponnesus, several data have been provided by systematic surveys (Flemming et al., 1973b). In addition, a curve of Holocene sea-level variations in the Kalamata Gulf has been obtained by a palaeogeographic study (Kraft et al., 1975).

The data available are summarized in Table 1. Submergence predominates almost everywhere, but varies in space and time. Indications on former shorelines have been deduced in some cases from submerged remains, the basal part of which is buried beneath an unknown depth of sediments. The submergence values given in Table 1, must therefore, be considered as minimum values. Concerning the ages ascribed to the remains, details on the criteria used for dating are seldom specified and it is possible that margins of uncertainty exceed in many cases several centuries.

On the whole, the distribution of data shows considerable changes in elevation even for remains from neighbouring localities. This suggests a strong fragmentation of the earth's crust, with many small blocks having different tectonic behaviours.

Submergence has been gradual in some cases and sudden at the time of earthquakes in other cases. On the site of Onugnatos, for instance, submergence seems to have been gradual. Several remains described by Flemming et al. (1973a, 1973b), between the coast of Elos Peninsula, west of Neapolis, and the islet of Pavlo Petri, are today submerged beneath 2 or 3m of water. The same site was on the surface at the time of Strabo and Pausanias, and was still fordable in 1677 (Baladié 1978,

Table 1: ARCHAEOLOGICAL INDICATORS OF SEA-LEVEL CHANGES IN THE NORTHWEST PART OF THE HELLENIC ARC

Locality				Type of remains	Present elevation (m) (- = below sea level)	Age (10^3 yrs)	References
N° (Fig. 2)	Name	Lat.N	Long.E				
1	Kalichiopulo	39°36'	19°54'	moles	-1.0 to -2.2	unknown	(1)
2	Leucadia (Levkas)	38°50'	20°43'	moles	-3.5	2.5	(1)
3	Phea (Katacolon)	37°38'	21°20'	ruins, quarries	≤-1.0	1.4	(2,4,6)
4	Pylus (Navarino)	36°57'	21°30'	ruins	-1.0/-1.5	2.0	(4,7)
5	Methoni	36°49'	21°42'	ruins	-1.5	1.0	(4,7)
6	Asine (Koroni)	36°48'	21°58'	ruins	-1.2	0.7	(4)
7	Messenia, Pamisos Valley	37°03'	22°05'	(9)	-2.0	2.0	(5)
8	Cardamyle (Kardamyli)	36°52'	22°15'	ruins	-1.0	1.0	(4,7)
9	Leuctra (Stoupa)	36°50'	22°16'	ruins	-2.5	2.0	(4,7)
10	Pephnus(Selenitsa)	36°49'	22°18'	ruins	-1.0	1.0	(4,7)
11	Tigani	36°32'	22°22'	tomb	-0.5	2.0	(4,7)
12	Taenarum (Limen Asomatos)	36°24'	22°29'	ruins	-1.0	2.3	(4)
13	Teuthrone (Kotronas)	36°37'	22°29'	ruins	-1.0	1.0	(4,7)
14	Skoutari	36°40'	22°30'	ruins	-3.5	1.5	(4,7)
15	Gytheum (Gythion)	36°45'	22°34'	ruins	-2.5	1.5	(1,4,7)
16	Trinasus (Trinisi)	36°47'	22°36'	ruins	-1.0	1.0	(4,7)
17	Acriae (Elea)	36°44'	22°48'	ruins	-2.0	3.0	(4)
18	Asopus (Plitra)	36°41'	22°50'	ruins	-2.0	1.5	(4)
18	" "	"	"	ruins	-3.0	3.0	(1,4)
19	Arkhangelos	36°37'	22°53'	quarries	-0.2	2.0	(4,7)
20	Onugnathus (Elaphonisos, Pavlo Petri)	36°28'	22°59'	fordable channel	≤-2.0	0.3	(4,6,8)
20	" "	"	"	ruins	≤-2.0	1.8	(4,6,8)
20	" "	"	"	ruins	≤-3.0	3.5	(1,3,4,7)
21	Boeae (Neapolis)	36°31'	23°02'	quarries	-0.3	1.5	(4,7)
22	Minoa (Monemvassia)	36°41	23°03'	quarries	-1.0	2.0	(4,7)
23	Zarax (Limen Ierakas)	36°47'	23°05'	quarries	-3.0	2.5	(4,7)

(1) Negris, 1904 ; (2) Hall, 1969 ; (3) Flemming et al., 1973a ; (4) Flemming et al., 1973b ; (5) Kraft et al., 1975 ; (6) Baladié, 1978 ; (7) Flemming, 1978 ; (8) the present study ; (9) paleogeographic reconstruction.

p. 295). At Phea, on the other hand, the subsidence movement was sudden. The town was very prosperous in late Roman times until the earthquake of 5 or 6 July 551 A.D., when it was partly engulfed by the sea (Baladié, 1978, p. 301).

Western Crete and Antikythira (Fig. 2, loc. 24 to 39).

The sequences of crustal movements inferred in western Crete and Antikythira during historical times from geomorphological, biological, petrological and radiometric evidence are not recorded directly in writings. Some historical and archaeological indications are nevertheless of help in estimating, by cross-checking, the age of these movements.

Ancient literature several times mentions the existence of harbours in western Crete, in localities where the coastal environment has been completely upset ever since the uplift. These harbours are: Dictynnaeum, Kisamos (at present Kastelli), Falasarna, Lissos, Suia, Peocilassus (at present Tripitis) and Tharra (at present Ayia Roumeli). These writings describe a situation which is necessarily earlier than the uplift. It is, therefore, interesting to try to establish when the ancient texts were written.

According to Muller (1965), Scylax of Caryanda, the explorer of Darios I, mentions "Phalasarna cum portu clauso" and "Lissa urbs cum portu" in his Periplus of the Interior Sea. According to other texts, however, the Periplus written by Scylax, of which only a few fragments are preserved, would have concerned only the coasts of Asia Minor from the Bosphorus to Cilicia in the Mediterranean. Pedech (1976), in particular, believes that the Periplus describing the Mediterranean basin ascribed to Scylax would, in fact, have been written later (Pseudo-Scylax) in the middle of the 4th Century B.C.E.

Strabo mentions several localities in Crete, including Phalasarna (X, 4, 2, X, 4, 13) and the Dictynnaeum sanctuary (X,4,12). It is believed that Strabo interrupted the writing of his Geography about 2 C.E. and that he resumed his work some 15 years later, in order to mention more recent events (Baladie, 1978). For Crete, however, the main sources of Strabo would have been Artemidor and Apollodorus, neither of whom, however, had ever been to the island (Lasserre, 1971).

If the reference to "Falasarna........clausum quae habeat portum" by Dionysus Calliphontis (Muller, 1965) does not add any new elements, that by Ptolemy (2nd century C.E.) on Falasarna, Lissos, Tharra and Peocilasium, in his Geography, indicates that these harbours were still functioning at that time.

The Stadiasmus, lastly, mentions among other ports, Tharram, Poecilassum, Suia, Lissum, Falasarna ("ibi statio est, emporium, urbs antiqua"), Cisamum and Dictynnaeo. The time when the Stadiasmus was written is controversial. Several authors ascribe the Stadiasmus to the beginning of our era. This is due, according to Pedech (1976), to the fact that it borrowed much from a Periplus of the Interior Sea, now lost, compiled by Menippus of Pergamum between 35 and 25 B.C.E. Rouge (1966) believes that the Stadiasmus was written slightly before the time of the Severus dynasty (A.D. 193-235); Pedech (1976) dates it at the 3rd century C.E.; finally, Pendlebury (1939) ascribes it to Byzantine times, between the 6th and the 11th century, adding (p. 24) that it must be the 6th century "for it mentions Phalasarna, which went out of use in the middle of the sixth century". The sources of this surprising, capital information, which would be consistent with the radiometric data available, are, however, not given.

In conclusion, ancient writings all seem to have borrowed much information from earlier Peripli. Sometimes they mention ports which were no longer used at the time these were written, whereas other ports, just constructed, were disregarded. This makes it hazardous to ascribe to these writings an absolute chronological value. However, until at least the 3rd century C.E., all the texts mention ports in western Crete that it would have been impossible to use after the uplift movement.

For the sake of comparison, the following situations are found today in the field:

Dictyna. Although the uplift movement here was about 3.5m, the site, which seems to have been located far away from towns and harbours, is still relatively sheltered and it is still easy to run aground with a boat (Flemming and Pirazzoli, 1981).

Kisamos (Kastelli). Remains of the ancient harbour have already been identified by several authors, including Spratt (1865) and Raulin (1869). The most prominent feature is the mole of rubble blocks, covered by marine incrustations, which rises several metres above sea-level, delimiting a small basin completely filled in with sand (Flemming and Pirazzoli, 1981). Nearby traces of marine corrosion indicate an uplift of about 6.5m.

Falasarna. The location of the ancient harbour was recognized by Spratt (1865), although it is now completely on the surface and there is no shelter along the coast. Rock surfaces have been cut by man, down to the level of a former shoreline, about 6.6m above sea-level, which is clearly marked by marine incrustations and sea-corrosion marks.

Very recently, an archaeological excavation has confirmed that the artificially excavated harbour, measuring 100 m x 75 m, was encircled by the city walls and that the uplift did not start earlier than the 4th century B.C.E. (Hadjidaki, 1986).

South of the ancient harbour a coastal quarrying site can be seen. Its floor was cut down to +5.8 m in altitude, i.e. lower than the uppermost shoreline marks, but seawards the rock was left in place at +7 m at least forming a barrier, so that the quarry's floor was not flooded.

Further to the west, a square tank was cut into the limestone very near the former shoreline. Two rock-cut staircases lead to the floor of the tank, at about +5.7 m, which is surrounded by slightly sloping benches (Fig. 3). According to Savignoni (1901), this is a salt pan. Cayeux (1907) has interpreted the basin as being a funeral chamber. Flemming et al. (1973a) suggest that it could have been a tank for the storage of water or other goods, more especially as the absence of opening towards the sea makes identification as a fish tank problematical. However, when the wind was blowing from the north or northwest, the site was very exposed to waves, and splashes could easily penetrate the basin. This makes the funeral and storage interpretations unlikely.

A Roman fish tank at Koutsounari, near Ierapetra (Davaras, 1978; Flemming and Pirazzoli, 1981), is very similar to the Falasarna basin: the same manner of cutting, the same type of staircases, the same benches projecting over the basin floor. The main difference consists of the two openings towards the sea which have been cut in Ierapetra, but do not exist in Falasarna. However, if the sea-level could rise again to the level of the ancient shoreline, which is well marked on the outer wall of the Falasarna tank, the floor of the basin would be rapidly flooded. As a matter of fact, a fissure several millimetres wide cuts across the seaward wall of the basin. A possible interpretation is that the fissure was opened by an earthquake, allowing sea water to come in, when the rock-cut basin, a fish tank on the model of that in Ierapetra, was still under construction. This made it unnecessary to cut openings, because the fissure was sufficient to allow a minimum water exchange with the sea, counterbalancing evaporation and draining wave splashes.

Lissos and Suia. This is the description given by Spratt (1865, 2: 240-242): "Lissos and Suia were proximate, the latter being the port of Elyros; but both, according to the 'Stadiasmus', possessed harbours, yet at neither is there now any place to shelter a boat unless ashore; but their positions are clearly recognizable by imagining a submergence to the extent indicated by the sea-marks upon the coast cliffs; and at Suia, a long line of ruins of buildings lies parallel to the shingle shore, which appears to have been a sort of embankment or mole.....of considerable width and length, that would be just above the sea at its old level previous to the upheaval and would render the level space lying behind it a well-sheltered and capacious port if cleared of the soil that has been washed into it."

The description is still valid now, with the difference that there are some new buildings and that a concrete pier has been constructed at Suia below the raised shorelines, which reach +6.6 m here.

Near Suia, Flemming (1978, Appendix I) reports, from indirect sources, that buildings of the 5th-8th century C.E. are now at +3 to +4 m, i.e. below the highest shoreline of historic age. When these constructions were built, the uplift movement had already occurred.

Poecilassus (Tripiti) Today the coast does not show any shelter for landing. Several levels of Pleistocene shorelines can be distinguished

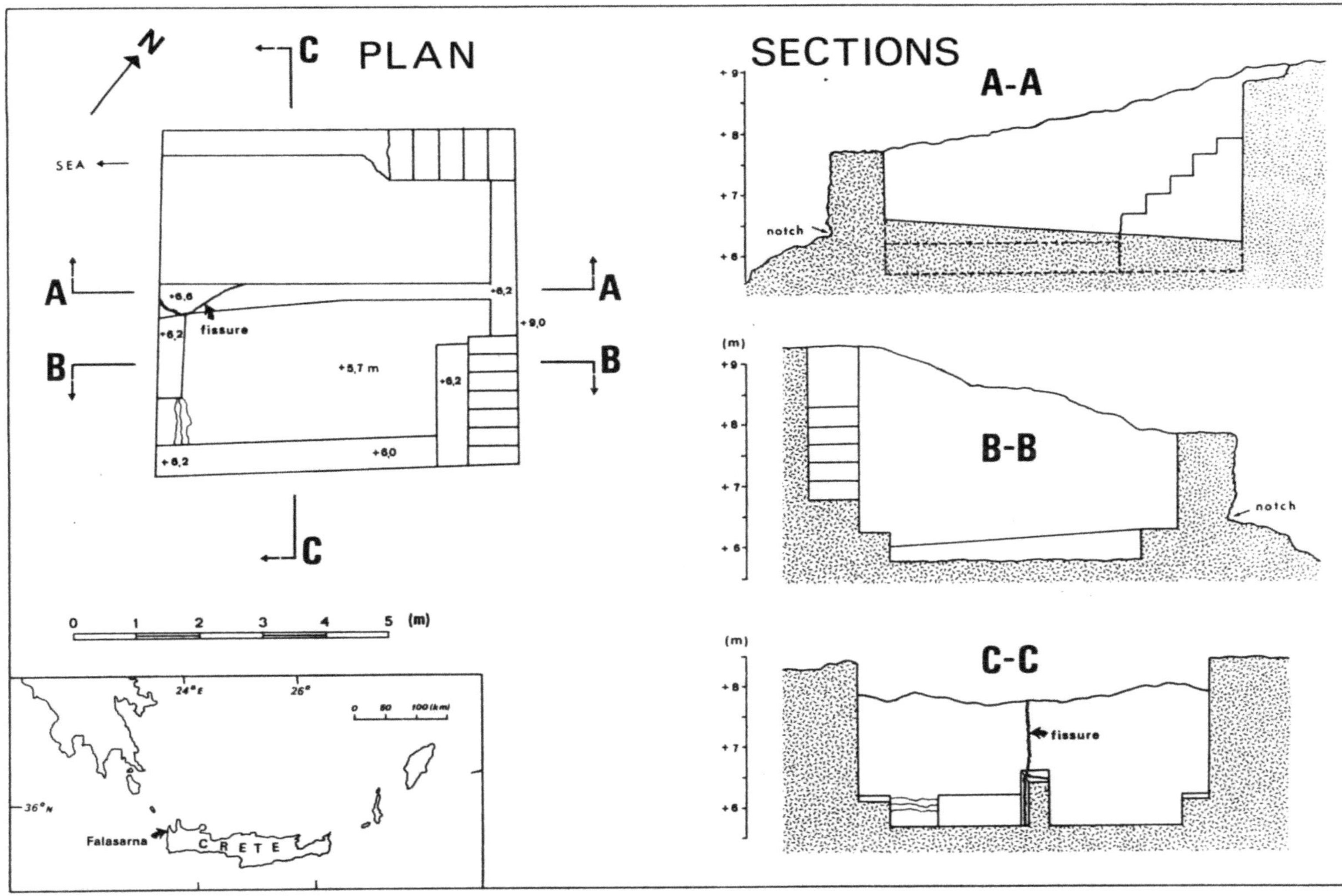

Figure 3: The Falasarna (fish) tank. Elevations are related to the sea level of 25 September 1979 at 1h P.M., with an accuracy of ± 0.1 m.

on the limestone cliffs which are undercut near the strand by many deep sea caves the entrances of which, polished by marine abrasion, are about 2 m above sea-level. At about +6.5 to +7.0 m, marks of recent marine biocorrosion are clear. When the sea was at this level, a narrow passage for boats existed between the vertical cliffs in the outlet of the Tripiti gorge, making it possible to reach a narrow but sheltered basin, expanding immediately within its entrance and forming a natural harbour.

Tharra (Ayia Roumeli). The aspect of this site, located at the outlet of the Samaria gorge, was probably not much altered by the uplift movement, which reached about 4.5 m here. It is still possible to find good anchorage in the embayment and, if the sea becomes rough, to ground boats on the beach.

An historical event concerning this area is the voyage of St. Paul along the coasts of Crete in 60 A.D. (New Testament, St. Luke, Acts of the Apostles, XXVII). Leaving from Fair Havens (Kaloi Limenes-nearby, behind Megalo Nisi, the Ay. Pavlos chapel was constructed to commemorate the passage of the Apostle) towards Phoenix (Loutron), the vessel was suddenly drawn out to sea by the wind: the voyage will end with a ship wreck in Malta. This description however does not provide any useful information on the coastline changes which occurred subsequently in Crete, since the harbours of Kaloi Limenes and Loutron are still in use today, as well as the small Bay of St. Paul near Lindos, in Rhodes, where the Apostle also put into port.

Several Venetian texts describe the damage caused by earthquakes in Crete, but not one of them mentions uplift movements in the western part of the island, which would certainly have endangered access to several fortified Venetian harbours (Canea, S. Teodoro, Castelli, Grabusa, etc.)

Although the western part of Crete has never been densely populated, it is very surprising that such exceptional tectonic movements, which must have upset life in the island, did not leave any trace in human memory or in local oral tradition. On this point Spratt (1865, 2, p. 230) is categorical: the uplift movement, although it dates from historical times, "is wholly without tradition or record".

Antikythira, which has followed the movements of Western Crete, is a small steeply sloping, infertile island, which has been deserted and repopulated more than once during its history. When the Florentine, Cristoforo Buondelmonti, visited this area, between A.D. 1414 and 1420, he reports (Legrand, 1897, p. 174) that the island, called Sikili, "was once inhabited, but today the only population consists of a crowd of wild asses". As the last repeopling of Antikythira seems to have come from western Crete, it is puzzling that Spratt (p. 231) states that the upheaval of Antikythira would date from historical times according to "local evidences and traditions".

In conclusion, according to all historical data available, the uplift of western Crete and the Antikythira area occurred no earlier than the 3rd century C.E. and before the arrival of the Venetians (13th century). According to the archaeological evidence, the uplift movement took place later than Roman times and earlier than the 5th-8th century A.D. (Suia). These conclusions are fully consistent with the radiometric data available.

The archaeological sea-level indicators available in western Crete, Antikythira and Gavdos are summarized in Table 2.

The masonry remains reported by Kelletat (1979: 44) from Yeoryioupolis are situated more than half a metre below the elevated shoreline which has been dated 2030± 70 yr BP(MC-2100) (Laborel et al., 1979). If the

Table 2: ARCHAEOLOGICAL INDICATORS OF SEA-LEVEL CHANGES IN ANTIKYTHIRA AND IN WESTERN CRETE

Locality				Type of remains	Present elevation (m) (- = below sea level)	Age (10^3 yrs)	References
N° (Fig. 2)	Name	Lat.N	Long.E				
24	Antikythira	35°53'	23°18'	slipway	≥+2.0	< 1.6	(9,10)
25	Panormon	35°25'	24°42'	unspecified	-0.5	0.5	(6)
26	Rethimnon	35°22'	24°28'	slipways	-0.2	2.2	(6,10)
27	Amfimala (Yeoryioupolis)	35°21'	24°16'	ruins	+0.9	unknown	(7)
28	Stavros	35°35'	24°06'	quarries	< 0	unknown	(7)
29	Dictyna (Menies)	35°40'	23°46'	religious site	+3.0	< 1.6	(5,10)
30	Kisamos (Kastelli)	35°30'	23°38'	moles	<+6.5	< 1.6	(1,2,10)
31	Falasarna	35°30'	23°34'	harbour	+6.6	< 1.6	(1,5,6,8, 10,11,13)
32	Moni Khrisos-kalitisas	35°19'	23°32'	sherds	+6.7	Minoan	(3,11,14)
33	Bienon (Krio)	35°13'	23°34'	harbour	+8.8 (max.)	< 1.6	(1,10)
34	Plakaki Kriou	35°13'	23°35'	sherds	>+7.7	Minoan	(4,14)
35	Lissos (Panayia)	35°14'	23°47'	harbour	>0(not measured)	< 1.6	(1,14)
36	Suia (Souyia)	35°14'	23°48'	harbour	+6.6 (max.)	<1.6	(1,14)
37	Peocilassus (Tripiti)	35°14'	23°52'	harbour	+6.5/+7.0 (max.)	<1.6	(1,14)
38	Tharrha (Ayia Roumeli)	35°13'	23°58'	harbour	+4.3 (max.)	<1.6	(1,14)
39	Gavdos, Cape Ay. Ioannis	34°51'	24°04'	sherds	+1.5/+2.5	unknown	(12)

(1) Spratt, 1865 ; (2) Raulin, 1869 ; (3) Hood, 1965 ; (4) Hood, 1967 ; (5) Flemming et al., 1973a ; (6) Flemming, 1978 ; (7) Kelletat, 1979 ; (8) Laborel et al., 1979 ; (9) Pirazzoli et al., 1981 ; (10) Flemming and Pirazzoli, 1981 ; (11) Thommeret et al., 1981 ; (12) Angelier, pers. comm. ; (13) Hadjidaki, 1986 ; (14) the present study.

sequence of vertical movements which have occurred in western Crete is taken into account, and if the masonry remains are in situ, their age can be either earlier than the 3rd century B.C.E., or later than the uplift, dated 1530±40 yr B.P.

The slightly submerged quarries at Stavros, showing a corrosion notch at +0.6 m on their walls (Kelletat, 1979: 47), indicate a former sea-level position which is not consistent with the general crustal movements in western Crete. This suggests that local tectonic phenomena are likely to have occurred in this locality, which seems situated along a fault (Angelier and Bonneau, 1976).

In Gavdos, marks of recent emergence are provided at +1.5/+2.5 m., by a calcareous sandstone, capping a sand layer with pebbles, which includes sherds of an unknown age. Gavdos Island is separated from Crete by a network of active faults (Nesteroff et al., 1977) and this means that any correlation with raised shorelines in Crete is uncertain.

Central Crete (Fig. 2, loc. 40 to 50).

Most archaeological remains available in central Crete give evidence of submergence (Table 3) and indicate minimum sea-level changes.

At Matala a slipway seems to be at the present sea-level, whereas tombs cut into the cliffs are found as much as 2 m underwater. Interpretations diverge.

According to Flemming (1978) the slipway would date from about A.D. 100 and the tombs from the Middle-Ages (A.D. 500 to 1000). Following this interpretation a land uplift movement would have occurred after the construction of the slipway, making the cutting of the tombs possible. Later, a subsidence movement of about the same amount would have brought back the slipway to nearly its former position. On the other hand, according to Evans (1923), the tombs should date from Greek times. In this case, their submergence would have occurred before the construction of the slipway and there would have been no noticeable change in the relative sea-level since that time. This interpretation is consistent with the absence of submergence traces at Kaloi Limenes.

Tombs at Lissos, similar to those at Matala, have been ascribed to the end of the 1st-2nd centuries C.E. (Tzedakis, pers. comm.). Lastly, Faure (1973, p. 44) has observed walls partly buried by sediments and sherds of many jars about 7 m deep in the middle of Matala Bay.

The sequence of vertical movements in Matala has still to be clarified in detail. In any case, these movements, whatever they may be, differ from those which occurred in the western part of the island. The Matala area is therefore located on a crustal block different from that of western Crete.

Table 3: ARCHAEOLOGICAL INDICATORS OF SEA-LEVEL CHANGES IN THE CENTRAL PART OF CRETE

N° (Fig. 2)	Locality: Name	Lat.N	Long.E	Type of remains	Present elevation (m) (- = below sea level)	Age (10^3 yrs)	References
	North Coast						
40	Minoa (Pirgos)	35°07'	25°44'	ruins	-1.5	2.2	(1,11)
41	Olous (Elounda)	35°16'	25°45'	ruins	-1.9	2.0	(2,11)
42	Port Kolokithia	35°16'	25°46'	unspecified	-1.0	1.0	(11)
43	Spinalonga	35°18'	25°45'	unspecified	-1.0	2.0	(11)
44	"	"	"	ruins	-2.4	2,6	(1,7)
44	Malia (Mallia)	35°18'	25°30'	ruins	-2.0	unknown	(12)
44	"	"	"	ruins	-1.0	3.6	(6,11)
45	Khersonisos	35°19'	25°24'	fish tanks	-1.0	1.9	(4,11)
46	Nirou Khani	35°20'	25°16'	ruins	-1.8	3.6	(3,7,11,13)
47	Amnisos	35°20'	25°13'	ruins	-1.5	3.6	(11,13)
	South Coast						
48	Moni Arvi	34°59'	25°28'	unspecified	-0.25	0.5	(11)
49	Lasea (Kaloi Limenes, Nisos Trafos)	34°56'	24°50'	harbour	±0	1.8	(5,10,11,13)
50	Matala	34°59'	24°45'	ruins	-7	unknown	(9)
50	"	"	"	slipway	±0	1.9	(11,13)
50	"	"	"	tombs	-2.0	1.1	(11,13)
50	"	"	"	tombs	-2.0	1.9/2.4	(3,8,14)

(1) Spratt, 1865 ; (2) Negris, 1904 ; (3) Evans, 1923 ; (4) Leatham and Hood, 1958 ; (5) Blanc, 1958 ; (6) Ecole Franç. d'Athènes, 1963 ; (7) Hafemann, 1965 ; (8) Bonnefont, 1972 ; (9) Faure, 1973 ; (10) Blackman and Branigan, 1975 ; (11) Flemming, 1978 ; (12) Guest Papamanoli and Treuil, 1979 ; (13) Flemming and Pirazzoli, 1981 ; (14) the present study.

On the north coast submergence does not increase gradually from west to east. This is evidence, contrary to the opinion of Raulin (1869), that the rigid tilting of western Crete did not extend to the central part of the island.

Eastern Crete (Fig. 2, loc. 51 to 63).

Several archaeological sea-level indicators have been reported from east Crete (Table 4). On the south coast an overall impression of relative sea-level stability is given by a tidal notch, well developed near the present sea-level, and by a Roman fish tank adjusted at a sea-level of - 0.2 m at Koutsounari, and by material of various ages (sherds of Minoan and historical times, a 1926 (A.D.) coin and a bicycle chain) (Dermitzakis and Theodoropoulos, 1975).

On the north coast a slight submergence predominates. On the east coast submergence is clear, increasing towards the north. However the pattern of a relative sea-level rise is not the same everywhere. At Mokhlos and Palaiokastron, for instance, sea-level indicators at various depths suggest a gradual subsidence movement. At Kato Zakros, on the other hand, a sudden subsidence jerk of about 1 m seems to have occurred in post-Roman times, following a period of stability in the relative sea-level which lasted at least 1500 years (Pirazzoli, 1980; Flemming and Pirazzoli, 1981).

Table 4: ARCHAEOLOGICAL INDICATORS OF SEA-LEVEL CHANGES IN EASTERN CRETE

Locality				Type of remains	Present elevation (m) (- = below sea level)	Age (10^3 yrs)	References
N° (Fig. 2)	Name	Lat.N	Long.E				
	South Coast						
51	Gralivia	35°01'	25°42'	sherds	<0 ?	3.8/4.5	(6)
52	Ierapetra	35°00	25°44'	unspecified	-0.5	1.8	(8)
53	Koutsounari	35°00	25°50'	fish tank, notch	-0.2	1.9	(7,9,12)
	Lesser Island						
54	Strongilo	34°57'	26°09'	notch	+0.9	2.2±0.1	(10,13)
54	"	"	"	notch	+0.5/+0.6	unknown	(10,13)
55	Koufonisi	34°56'	26°10'	notch	+1.0	1.9	(4,12,14)
	East Coast						
56	Xirokambos	35°03'	26°16'	salt marshes?	+0.3	unknown	(13)
57	Kato Zakros	35°06'	26°17'	fish tank ? notch	-1.0	1.9 ?	(9,11,12)
57	" "	"	"	bathing pool, notch	-1.0	3.6	(11,12)
58	Cape Plaka	35°11'	26°20'	unspecified	-0.2	1.0	(8)
59	Palaiokastron, Kastri	35°12'	26°18'	beachrock	-0.5	unknown	(11)
59	"	"	"	kiln	-2.0	1.6/2.2	(11,12)
60	Itanos, Erimopolis	35°16'	26°17'	ruins	-2.2	2.0	(1,8)
	North Coast						
61	Sitia	35°13'	26°08'	basins	+0.3	1.9	(5,9)
61	"	35°12'	26°07'	unspecified	-0.8	1.8	(8)
62	Mokhlos	35°11'	25°55'	fish tank	-0.6/-0.7	1.9	(2,9,12)
62	"	35°09'	25°53'	sherds	-1.75	3.6	(3,8)
63	Psira	35°11'	25°52'	unspecified	-0.75	1.9	(8)

(1) Spratt, 1865 ; (2) Leatham and Hood, 1959 ; (3) Hafemann, 1965 ; (4) Flemming et al., 1973a ; (5) Davaras, 1974 ; (6) Dermitzakis and Theodoropoulos, 1975 ; (7) Davaras, 1978 ; (8) Flemming, 1978 ; (9) Pirazzoli, 1979b ; (10) Pirazzoli, 1979a ; (11) Pirazzoli, 1980 ; (12) Flemming and Pirazzoli, 1981 ; (13) Montaggioni et al., 1981 ; (14) Peters, 1985.

On the islet of Strongilo three ripple notches have developed at + 0.9 m, at + 0.5 to 0.6 m and at the present sea-level (Pirazzoli, 1979a, Fig. 3). The shoreline at + 0.9 m has been dated 2200± 100 yr BP (Montaggioni et al., 1981). On the nearby Koufonisi Island emergence since Roman times was estimated at about 2 m by Flemming et al. (1973), but seems in fact to be only 1 m (Peters, 1985), as in Strongilo.

Karpathos (Fig. 2, loc. 64 to 68).

Holocene emerged shorelines seem to be completely absent in Karpathos, where several underwater shorelines indicate recent subsidence movements with vertical displacements of at least 1.0 to 1.5 m, dating from historical times. The amount of submergence does not change significantly from place to place, suggesting the absence of a tilting motion.

At Iardani (this same locality has been erroneously called Vathi Potamos in some previous publications), a series of five superimposed shorelines, three of which are cut into the walls of a submerged quarry, indicate that changes in the relative sea-level have occurred by sudden jerks (Flemming, 1978; Pirazzoli, 1980). The last jerk was recent enough, for no notch has formed yet at the present sea-level.

Table 5: ARCHAEOLOGICAL AND MORPHOLOGICAL INDICATORS OF SEA-LEVEL CHANGES IN KARPATHOS

Locality				Type of remains	Present elevation (m) (- = below sea level)	Age (10^3 yrs)	References
N° (Fig. 2)	Name	Lat.N	Long.E				
64	Vrucunda (Vurgunda)	35°48'	27°12'	mole	-1.0 ?	2.0	(1,2)
64	"	"	"	notch	-1.0	2.0 ?	(1)
65	Lefcos	35°35'	27°06'	ruins	-1.0	2.0	(1,2)
66	Kyra (Kera) Panagia	35°35'	27°13'	notch	-1.1/-1.2	unknown	(3)
67	Iardani (Vathi Potamos, Amorfo Bay)	35°29'	27°14'	quarries	-1.0	historical	(1,4)
67	" "	"	"	notch	-0.2	"	(4,1)
67	" "	"	"	notch	-0.55	"	(4,1)
67	" "	"	"	notch,platform	-0.7	"	(4,1)
67	" "	"	"	notch,platform	-1.5	historical?	(4)
67	" "	"	"	platform	-3.0	unknown	(4)
68	Makriyalo	35°26'	27°11'	slipway	±0 ?	0.6 to 1.0	(1)
68	"	"	"	notch	-0.9	unknown	(1)
68	Nicolachi Point	"	"	platform	-1.0	unknown	(5)

(1) Flemming, 1978 ; (2) Barrier, 1979 ; (3) Pirazzoli, 1979a ; (4) Pirazzoli, 1980 ; (5) the present study.

Table 6: ARCHAEOLOGICAL INDICATORS OF SEA-LEVEL CHANGES IN RHODES

Locality				Type of remains	Present elevation (m) (- = below sea level)	Age (10^3 yrs)	References
N° (Fig. 2)	Name	Lat.N	Long.E				
69	Rhodes, Campo golf	36°25'	28°17'	sherds (in a marine incrustation)	+0.5	Roman times	(1,6,7)
69	between Rhodes	36°26'	28°17'	quarries	-0.4/-0.5	Roman times	(4,6,7)
70	and Calitea	36°22'	28°17'				
71	Faliraki	36°20'	28°15'	sherds (in a marine incrustation)	+1.0	unknown	(5)
72	Ladico	36°19'	28°15'	sherds (in a marine incrustation)	+1.0	1.6/2.3	(2)
73	Caraci	36°10'	28°08'	sherds (in a marine incrustation)	+0.3	2.3/2.6	(6)
74	Lacania	35°57'	27°56'	sherds (in a marine incrustation)	+0.3	2.3/2.5	(6)
75	Angonia (Kamiros Skala)	36°16'	27°52'	ruins	>0	1.4	(3,7)
75	" "	"	"	"	-2.0	unknown	(7)

(1) Mutti et al., 1970 ; (2) Dermitzakis and Theodoropoulos, 1975 ; (3) Marinos and Dermitzakis, 1977 ; (4) Flemming, 1985 ; (5) Gauthier, 1979 ; (6) Pirazzoli et al., 1981, 1982b ; (7) the present study.

The main characteristic of the sea-level data available in Karpathos are summarized in Table 5. Underwater photographs of submerged features have already been published (Pirazzoli, 1979a, Fig. 4; 1980, Fig. 7 to 10).

Rhodes (Fig. 2, loc. 69 to 75).

Archaeological remains in relation to sea-level consist mostly of slightly submerged coastal quarries (especially between the town of Rhodes and Calitea (Kallithea), of sherds included in exposed marine incrustations on the east coast, and of submerged ruins near Kamiros Skala on the west coast (Table 6).

Near Cape Vudhi and Calitea, where the floor of wide quarries is slightly underwater, almost horizontal scattered surfaces have been left at various heights above sea-level. Flemming (1978, Appendix 1) interprets surfaces at about +2.9 m as the floor of quarries cut near a former sea-level during Greek times, but the surfaces may also correspond to the base of certain layers of rock preferred for quarrying.

The coastal quarries of Rhodes have been ascribed by Flemming to the 4th century B.C.E. and/or to the 14th century C.E., the latter assumption being based on block sizes which seem to correlate with the fortifications built by the Knights of St. John. However, marine incrustations collected in situ at +2.35 m in altitude have been dated 2280 ± 110 yr B.P. (Pa-447), indicating that during Greek times the relative sea-level was much higher than at present in this area. It was probably only in 228 B.C.E., during the great earthquake that destroyed the Colossus, that a sudden uplift movement caused the emergence of the limestone formations which were later quarried on the coast. During occupation by the Knights, the sea-level was again higher than at present, as proved by fossil barnacles collected in a growth position near Calitea and dated 855 + 70 yr BP (Pa-418) at + 1.1 ± 0.3 m and 390 ± 70 yr BP (Pa-416) at 1.5 ± 0.2 m (Pirazzoli et al., in preparation). This means that the coastal quarries between Rhodes and Calitea could have been used only during Roman times.

At Kamiros Skala the remains of the Byzantine church mentioned by Dermitzakis and Theodoropoulos (1975) cannot be considered as evidence of submergence. The lowest part of the church's foundations which are exposed near the beach have been measured by this author at +0.55 m and there is a mosaic floor at +1.8 m. The current destruction of the church ruins is due essentially to marine erosion. The nearby shallow sea floor is encumbered with reworked blocks, probably having fallen from the church. More seawards, however, traces of a foundation wall disappearing beneath the sand have been found at about -2 m. However, this structure does not belong to the church and its age is unknown.

DISCUSSION AND CONCLUSION

In western Crete and Antikythira the data available are consistent with the Early Byzantine Tectonic Paroxysm which has been inferred from geomorphological, biological, petrologic and radiometric data all around the eastern Mediterranean basin, occurring about 1530 ± 40 yr BP (Pirazzoli, 1986). According to conversion tables, this radiometric age may correspond to any calendric date between C.E. 430 and 580. Historical and archaeological data do not at the moment furnish more precise dates, but confirm that the uplift movement which raised western Crete occurred later than the 3rd century C.E. and before the 5th-8th century C.E.

Up to the present, archaeological investigations have focussed mostly on the main occupation periods of sites. Very few scholars have attempted to date the end of occupation periods. Nevertheless, more information on the latter point would be of paramount interest in several coastal sites of the eastern Mediterranean.

Among several major earthquakes which may have accompanied the EBTP (see Pirazzoli, 1986, Table 1), that of July 551 A.D. is known to have caused drastic damage along all the Levant coasts (Brown, 1969). It is interesting to note that at least one site in the Peloponnesus (Phea, loc. 3), located a few dozen miles from the Hellenic Trench, was obviously affected at exactly the same time. This may be due to a subduction thrust of the Ionian oceanic crust under the Hellenic arc.

Unfortunately, similar historical data concerning Crete are lacking. At any rate, the date of 551 C.E. would be consistent not only with the radiometric data obtained in western Crete and Antikythira, but also with the unreferenced statement by Pendlebury (1939) that Falasarna harbour ceased activity in the middle of the sixth century C.E.

It is essential to improve the data available and prepare a more complete and detailed inventory of well-dated vertical movements which have occurred in coastal areas of the eastern Mediterranean during the last few thousand years. The aim is to define more precisely the eventual effects of the EBTP and those of its preparatory phases in all the areas which may have been affected. An inventory of this type is needed to obtain a better insight into the geodynamic phenomena which are deforming the eastern Mediterranean region.

ACKNOWLEDGEMENTS

I should like to thank Dr. R.R. Stieglitz (Rutgers University, Newark) for the very helpful suggestions and for having brought to my attention the publication by Brown (1969), Ms. M. Delahaye (Institut Océanographique, Paris) for revision of the English text and Ms. D. Maraine (CNRS-Intergéo, Paris) for typing and page-setting the manuscript.

REFERENCES

Angelier, J. 1979. Néotectonique de l'arc égéen. Soc. Géol. Nord, Publ. no. 3, 418 p.

Angelier, J. and M. Bonneau. 1976. Sur les enseignements structuraux des photographies transmises par satellite artificiel : l'exemple de l'ile de Crète (Grèce). C.R. Acad. Sc. Paris, t. 283, D. : 603-606.

Baladié, R. 1978. Strabon. Géographie. Tome V (Livre VIII). Texte établi et traduit. Les Belles Lettres. Paris. 317p.

Barrier, E. 1979. Etude Néotectonique des Iles de Karpathos et Kassos. Etude Géologique de l'Ile de Kassos (Dodecanèse ,Grèce). Thèse. Univ. P. et M. Curie, Dept. Géotectonique, 204 p.

Blakman, D.J. 1973. Evidence of sea-level change in ancient harbours and coastal installations. In: Marine Archaeology, Colston Papers, 23: 115-139.

Blackman, D.J. 1982a. Ancient harbours in the Mediterranean. Part 1. Int. Journ. Naut. Archaeol. 11: 79-104.

Blackman, D.J. 1982b. Ancient harbours in the Mediterranean. Part 2. Int. Journ. Naut. Archaeol. 11: 185- 212.

Blackman, D. and J. Branigan. 1975. An archaeological survey of the south coast of Crete between Ayiofarango and Chrisostomos. Ann. Brit. Sch. Athens 70: 17-36.

Blanc, J.J. 1958. Recherches géologiques et sédimentologiques. Campagne 1955 en Méditerranée Nord-Orientale. Rés. Sc. Camp. "Calypso" Ann. Inst. Océanogr. 34, fasc. 3: 157 - 211.

Bonnefont, J.C. 1972. La Crète: Etude Morphologique. Univ. Lille III, Service de reproduction des thèses, 845 p.

Brown, J.P. 1969. The Lebanon and Phoenicia. Ancient Texts Illustrating their Physical Geography and Native Industries. Vol. 1, p. 126 et seq.,Beirut.

Cayeux, L. 1907. Fixité du niveau de la Méditerranée, à l'époque historique (Délos, Crète). Ann. Géogr. XVI: 97 - 116.

Davaras, C. 1974. Rock-cut fish tanks in eastern Crete. Ann. Brit. Sch., Athens 69: 87-93 +pl.

Davaras, C. 1978. A rock-cut fish tank near Hierapetra (in Greek, with English abstract). Archaiologikon Deltion, 30 (1975): 149-157.

Dermitzakis, M. and D. Theodoropoulos. 1975. Study of beach rock in the Aegean Sea. Observations on occurrences in SE Crete, Rhodes and Metopi. (in Greek, with English abstract). Ann. Géol. Pays Hellen., 1, XXVI: 275-305.

Ecole Française d'Athènes. 1963. Fouilles executées à Mallia. Etude du site. Etudes Crétoises, XIII, 137 p. +pl.

Evans, A. 1923. Crete as a stepping-stone of early culture. Nature, vol. 112: 660-662.

Faure,P. 1973. La Vie Quotidienne en Crète au Temps de Minos. Hachette Littérature. 396 p.

Flemming, N.C. 1969. Archaeological Evidence for Eustatic Change of Sea-Level and Earth Movements in the Eastern Mediterranean During the Last 2000 Years. The Geological Society of America. Special Paper 109, VIII + 125 p.

Flemming, N.C. 1978. Holocene eustatic changes and coastal tectonics in the northeast Mediterranean : implications for models of crustal consumption. Phil. Trans. R. Soc. London, A, 289 (1362): 405-458 + appendix 1.

Flemming, N.C. 1979. Archaeological indicators of sea level. Oceanis, 5, Fasc. hors-serie : 149-166.

Flemming, N.C. Czartoryska, N.M.G. and P.M. Hunter. 1973a. Archaeological evidence for vertical earth movements in the region of the Aegean Island Arc. In: Science Diving International. Proc. 3rd Symp. Sci. Commit., Conf. Mond. Activ. Subaq., 8-9 Oct., London : 47-65.

Flemming, N.C., Czartoryska, N.M.G. and P.M. Hunter. 1973b. Archaeological evidence for eustatic and tectonic components of relative sea-level change in the South Aegean. In: Marine Archaeology. D.J. Blackman ed. Colston Papers, vol. XXIII : 1-66.

Flemming, N.C. and P.A. Pirazzoli. 1981. Archéologie des côtes de la Crète. Histoire et Archéologie. Les Dossiers no. 50 : 66-81.

Gauthier, A.J. 1979. Contribution à l'Etude Néotectonique du Domaine Egéen: l'île de Rhodes. Thèse, 3e cycle. Univ. Paris XI (Orsay) 207 p. + pl. + bibl.

Guest-Papamanoli, A. and R. Treuil. 1979. Malia. II. Bâtiment immergé. Bull. Corresp. Hellen. 103 : 668-669.

Hadjidaki, E. 1986. Excavations at the Classical/ Hellenistic harbour of Phalassarna, Western Crete, Greece. Cities on the Sea - Past and Present - 1st Intern. Symp. on Harbours, Port Cities and Coastal Topography, Haifa, Israel, September 22-29, 1986. (Oral Presentation).

Hafemann, D. 1965. Die Niveauveränderungen an den Küsten Kretas seit dem Altertum. Abhandlungen der Akademie der Wissenschaften und der Literatur. Mathematisch-Naturwissenschaftlichen Klasse. NR 12 : 605-688 + pl.

Hall, J.E. 1969. Ancient underwater city. Oceans, 1, 4: 48- 51.

Hood, M.S.F. 1965. Minoan sites in the far west of Crete. Ann. Brit. School at Athens, 60: 99-113 + pl.

Hood, M.S.F. 1967. Some ancient sites in South-west Crete. Ann. Brit. School at Athens, 62 : 47-56 + pl.

Issel, A. 1883. Le oscillazioni lente del suolo o bradisismi. Atti. Univ. Genova, 5, 422 p.

Kelletat, D. 1979. Geomorphologische Studien an den Küsten Kretas. Abh. Ak. Wiss. Göttingen, 32 : 1-105.

Kraft, J.C. Rapp, G. and S.E. Aschenbrenner. 1975. Late Holocene paleo-geography of the coastal plain of the Gulf of Messenia, Greece, and its relationships to archaeological settings and coastal change. Geol. Soc. Amer. Bull. 86 : 1191-1208.

Laborel, J. Pirazzoli, P.A., Thommeret, J. and Y. Thommeret. 1979. Holocene raised shorelines in western Crete (Greece). Proc. "1978 Intern. Symp. Coastal Evolution in the Quaternary". Sâo Paulo : 475-501.

Lasserre, F. 1971. Strabon, Géographie, Tome VII (Livre X). Texte établi et traduit. Les Belles Lettres, Paris, 170 p.

Leathan, J. and S. Hood. 1958-1959. Submarine exploration in Crete, 1955. Annual British School at Athens, 53-54: 263-280 + pl.

Legrand, E. 1897. Description des Iles de l'Archipel, par Christophe Buondelmonti. Traduction francaise et commentaire d'une version grecque par un anonyme. Leroux. Paris. 260 p.

Leycester, E.M. 1852. Some account on the Volcanic Group of Milo, Anti-Milo, Kimolo and Polino. Geog. Soc. J., London, t. 22 : 201-227.

Marinos, G. and M. Dermitzakis. 1977. Contribution to the study of coastal geology of Rhodes island (Greece). Rapp. Comm. Int. expl. sc. Mer Mediterranee, 24, 7a: 271-273.

Montaggioni, L.F., Pirazzoli, P.A., Laborel, J., Thommeret J. and Y. Thommeret. 1981. Rivages tyrrhéniens et historiques à Strongilo et dans le sud-est de la Crète (Grèce). Implications néotectoniques. In : Actes Coll. "Niveaux Marins et Tectonique Quaternaires dans l'Aire Méditerranéenne". CNRS et Univ. Paris I : 67-76.

Müller, K. 1965. Geographi Graeci Minores. Georg. Olms Verslag., Hildesheim.

Mutti, E., Orombelli, G. and R, Pozzi. 1970. Geological studies on the Dodecanes Islands (Aegean Sea). IX : Geological map of the Island of Rhodes (Greece). Explanatory notes. Ann. Géol. Pays Helléniques, 22: 77-226.

Negris, P. 1904. Vestiges antiques submergés. Athen. Mitteilungen, 29 :340-363.

Nesteroff, W.D. Lort, J., Angelier, J. Bonneau, M. and A. Poisson. 1977. Esquisse structurale en Méditerranée orientale au front de l'arc égéen. Intern. Symp. Struct. Hist. Medit. Bas., Split, 1976, Technip, Paris: 241-256.

Pédech, P. 1976. La Géographie des Grecs. Presses Univ. France, Coll. SUP no. 5, 202 p.

Pendlebury, J.D.S., 1939. The Archaeology of Crete. Methuen, London, 400 p.

Peters, J.M. 1985. Neogene and Quaternary vertical tectonics in the south Hellenic arc and their effect on concurrent sedimentation processes. GUA Pap. Geol. 1, 23, 247 p.

Pirazolli, P.A. 1976. Les variations duniveau marin depuis 2000 ans. Mém. Lab. Géomorphol. Ec. Prat. Haut. Et. no. 30, 421 p. Dinard.

Pirazolli, P.A. 1979a. Encoches de corrosion marine dans l'arc hellénique. Oceanis, 5, Fasc. Hors-série : 327-334.

Pirazolli, P.A. 1979b. Les viviers à poissons Romains en Méditerranée. Oceanis, 5, Fasc. Hors-série : 191-201.

Pirazolli, P.A. 1980. Formes de corrosion marine et vestiges archéologiques submergés: interprétation néotectonique de quelques exemples en Grèce et en Yougoslavie. Ann. Inst. Océanogr. 56 (S): 101-111.

Pirazolli, P.A. 1986. The early Byzantine Tectonic Paroxysm. Zeitsch, für Geomorph., Suppl.--Bd. 62: 31-49.

Pirazolli, P.A., Thommeret, J., Thommeret, Y., Laborel, J. and L.F. Montaggioni. 1981. Les rivages émergés d'Antikythira (Cerigotto): corrélations avec la Crète occidentale et implications cinématiques et géodynamiques. In: Actes Coll. "Niveaux Marins et Tectonique Quaternaires dans l'Aire Méditerranéenne". CNRS et Univ. Paris I: 49-65.

Pirazzoli, P.A., Thommeret, J., Thommeret, Y., Laborel, J. and L.F. Montaggioni. 1982a. Crustal block movements from Holocene shorelines: Crete and Antikythira (Greece). Tectonophysics, 86: 27-43.

Pirazzoli, P.A., Montaggioni, L.F., Thommeret, J., Thommeret, Y. and J. Laborel. 1982b. Sur les lignes de rivage et la néotectonique à Rhodes (Grèce) a l'Holocène. Ann. Inst. Océanogr., 58, 1 : 89-102.

Raulin, V. 1869. Description Physique et Naturell de l'Isle de Crète. Paris, Arthus Bertrand, 3 t. 1078 p.

Rougé, J. 1966. Recherches sur l'Organisation du Commerce Maritime en Méditerranée sous l'Empire Romain, Paris.

Savignoni, L. 1901. Esplorazione archeologica delle province occidentale di Creta. Monumenti Antichi, 11, Sp.: 285-550.

Schmiedt, G. 1972. Il Livello Antico del Mar Tirreno. Olschki, Firenze, 323 p.

Schmiedt, G. 1981. En Italie, les viviers romains de la côte thyrrhénienne et la montée du niveau des eaux. Histoire et Archéologie. Les Dossiers, 50 : 28-45.

Spratt, T.S. 1865. Travels and Researches in Crete. London. 2 vol., 387 + 435.

Suess, E. 1888. Das Antliz der Erde (La face de la Terre). Trad. française par E. De Margerie. Tome 2. Colin, Paris, 1900. 878 p.

Thommeret J. and Y. Laborel, J. Montaggioni, L.F. and P.A. Pirazzoli. 1981. Late Holocene shoreline changes and seismo-tectonic displacements in western Crete (Greece). Zeits. f. Geomorphol., Suppl.-Bd. 40: 127-149.

Trousset, P. and R. Guery. 1981. Carrières antiques de la Couronne. Quatrième Centenaire de l'Union des Trois Quartiers de Martigue: 55-71.

AVNER RABAN

Coastal Processes and Ancient Harbour Engineering

The coast of the south-eastern Mediterranean is, in its present configuration, rather hostile to long shore nautical activity and lacks natural topographic features that could be used as harbours. These facts were well known to the ancients who had to discover technical ways of overcoming the problems that nature had posed on ways of utilizing the coast actively in the seaborne trade, either between Egypt and Levanto-Syrian port cities, or as the Mediterranean base for the overland transport of goods from the Red Sea and the Indian Ocean. It appears that generations of observations led the ancient harbour engineers to discover the proper technical answers to most of these problems in order either to maintain the functioning of old harbours through changes in the coastal topography, or to select new sites for building artificial harbours, so that they could be kept operational for many generations with minimal maintenance.

In order to point out some of these engineering measures we have chosen data collected in recent years from the ancient Herodian harbour of Caesarea as an example of the local traditions in harbour construction(1).

The Herodian harbour of Caesarea was known in antiquity as Sebastos (the Greek equivalent of the Latin epitaph Augustus - the Blessed, which was the chosen name of the first Roman emperor Octavian, Herod's patron). Sebastos not only brought about the building of Caesarea, but was also an exclusive administrative unit within the metropolis. The Jewish historian, Josephus, who lived half a century after the harbour was built, described its features and the way it was built with several direct references to the coastal oceanographic uniqueness of the site. He wrote:

> "This city, situated in Phoenicia, in the passage by sea to Egypt, Joppa and Dora, which are lesser maritime cities and not fit for havens, on between the account of the impetuous south winds that beat upon them, which rolling sands that come from the sea against the shores do not admit ships lying in their station; but the merchants are generally there forced to ride at their anchors in the sea itself." (Antiquities, XV: 335).

The recent studies of CAHEP(2) have revealed that the ancient engineers of Sebastos were well aware of these facts and would have taken proper measures. Such measures included:

I. BUILDING SEA WALLS ON SAND

It seems that from the very earliest stages of harbour engineering people learned to prevent the undertrenching of walls laid on sand by paving it with a layer of relatively large pebbles. The layer was thick enough and the stones of sufficiently large size so as not to be carried away by the current created at the foot of the sea wall during its heaviest surge. This was so in the case of the 7th-6th century B.C.E. Phoenician harbour of Athlit and the somewhat later seawall of the port of Akko(3) (Fig. 1).

In Caesarea the cushioning of the sandy sea floor was carried out only at the foot of the external sloping face of the great breakwater and under the base of the subsidiary one (see below and Figs. 2a-2b). It seems that the technique of cushioning the sandy sea floor was invented in the Levant by the Phoenicians, probably with the innovation of free-standing ashlar sea walls(4).

The undertrenching effect of wave-initiated currents on solid structures based on a sandy bottom was well-known to the ancient harbour engineers. In Rome it was Vitruvius who had suggested in the last century B.C.E. that:

> "Then, the sloping section is to be filled in with sand and brought up to the level of the retaining walls and platform surface. Next, a pier of appointed size is to be built there, on this levelled surface, and when it has been poured is left at least two months to dry. Then the retaining wall which holds in the sand is cut away, and in this manner erosion of the sand by the waves causes the pier to fall into the sea". (De Architectura, 5.12.4).

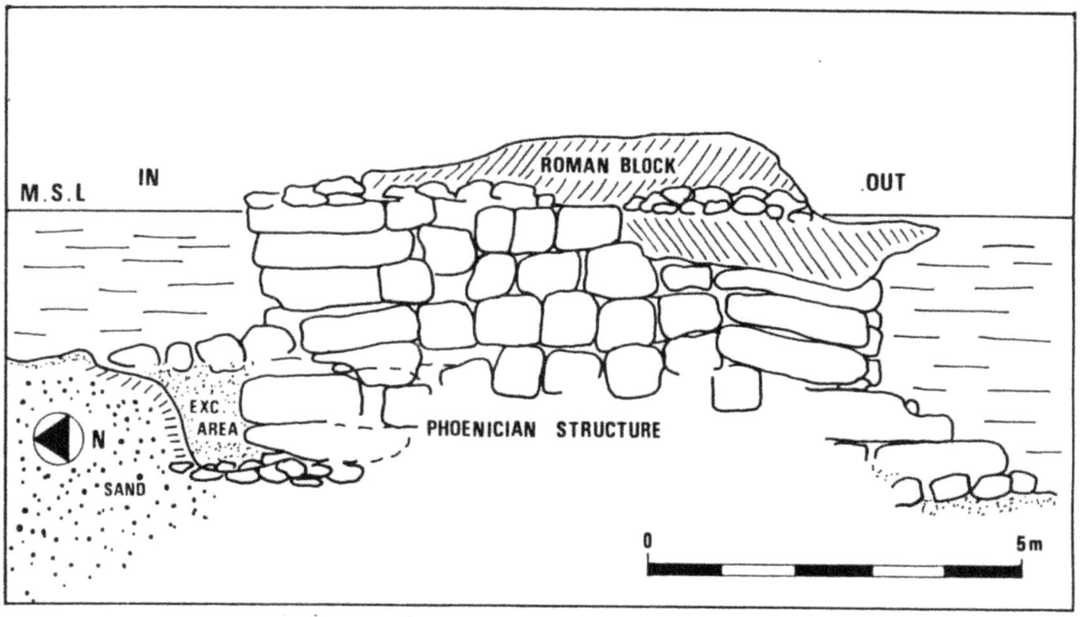

Figure 1: A cross-section of the Phoenician mole at Akko.

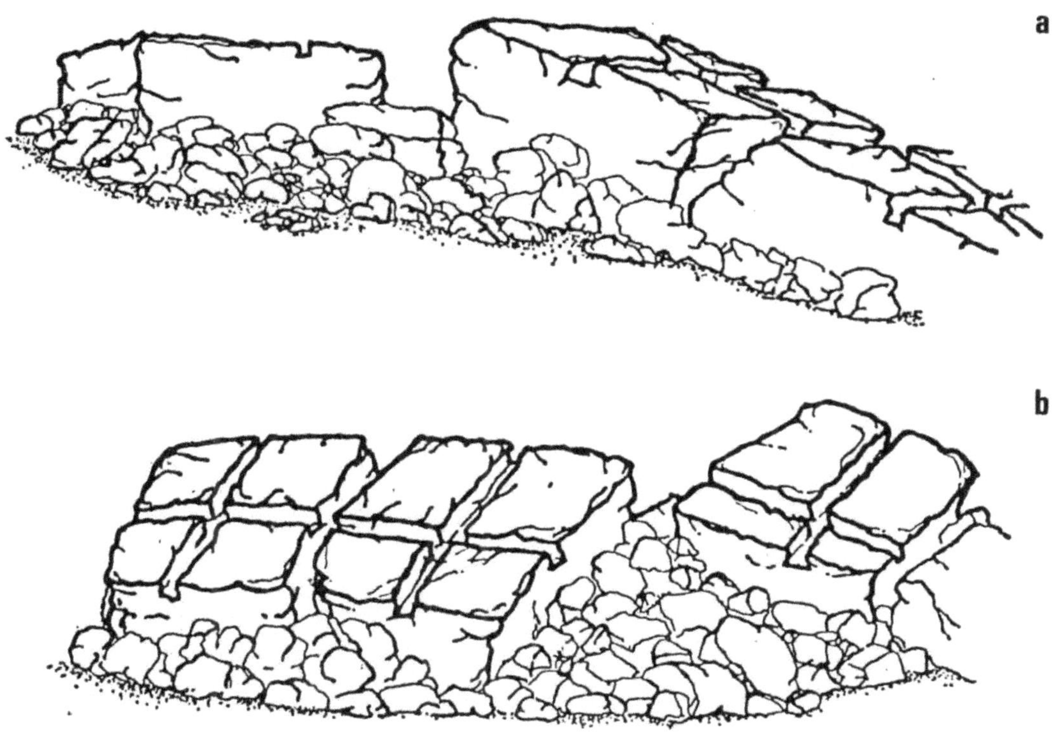

Figure 2: The rubble cushion of the breakwater at Caesarea. (L. Vann).

In this respect, it is quite interesting to learn about modern breakwaters along the Levant coast of the Mediterranean that have been built recently. They suffer from various rates of tilting and subsidence due to the undertrenching affect along their external base. The largest and most modern commercial harbour of Israel, that of Ashdod has suffered the repeated destruction of its main breakwater. This process was eventually stopped, but not before a large quantity of rubble was dumped on the sea floor along its exterior.

II. SUBMERGED CAISSONS

The underwater excavations of CAHEP during the 1982-83 seasons revealed an advanced and sophisticated adaptation of the recommended Vitruvian methods of building conglomerated sea walls(5). This is the first time that this well-known classical reference has been properly illustrated in an archaeological context, and found in such a state of preservation that the actual way it was used can be followed through the sequence of the stages of construction(6). Though the technique of using sea hardening cement was most probably of Italian origin, and at Caesarea represented a recently innovative and imported skill, its actual adaptation to local conditions seems to have been genuinely correct. The shifting sandy bottom of the sea at this site necessitated a certain modification of the three alternative methods which Vitruvius referred to. The fact that the location of Sebastos is exposed to the full surge of the sea storms was probably also a unique one for the harbours of the period.

Figure 3: The base beams for the wooden caissons at Caesarea. (Photo: M. Little).

For these reasons the caissons for the sea walls of the Herodian harbour at Caesarea had to be prefabricated on land, floated, and then towed toward the precise siting in complete units. Chained to the heavy sinkers on the sea bottom, the hollows between the double walls would be evenly filled with pozzolana mortar until its load would cause the caisson to sink slowly to its required resting place on the sea bottom. The combined structure of such caissons, reinforced by the ashlar and rubble-retaining mass, would secure the positioning of the frames for as long as it would take for the mixed fill of rubble and pozzolana to be agglomerated and hardened by the seawater. As there was no proper impregnated floor at the base of these caissoned blocks it is assumed that the fill was dumped into frames full of water without any attempt to pump off the sea. This would make the process of building much faster and easier, and though recommended by Vitruvius(7), it seems as if the pumping was not really necessary - as the case at Caesarea shows. (Figs. 3, 4).

III. WAVE CARRIED SAND TO FILL-IN MOLES

During the 1984 season of CAHEP it was decided to try and go down lower than what seemed to be a regular segment of the large breakwater surface - a somewhat levelled area of medium-sized rubble, covering the area between what might have been the spinal wall of the structure and the inner face of the mole. That dig, which was designated as D3 went down from the level of -5.2m below MSL having been exposed, just below the surface layer of rubble, over 3 metres of well-stratified and well-defined laminas of sand classified by grain size. Some inter-bedded thin layers contain pieces of wood, broken reeds and twigs(8) (Fig. 5). All through these sand layers neither pottery sherds nor shells were found. At a depth of -8.3 m there was another type of sand, with shells and some Late Hellenistic (Pre-Herodian) pottery sherds. Digging into that sand showed it to be non-stratified and similar to the present sandy seabottom.

From a sedimentological point of view it is apparent that the sandy fill between -5.3 - -8.3m below present MSL represents a final deposition that had occurred by variable wave energy and within a confined hollow. Based on a series of data which documented the rate of subsidence of the seabottom in that area since the Herodian period to somewhat over 5m(9) and the original depth of the harbour's floor to about 3m(10), it must be deduced that the sandy fill represents the very volume of the originally submerged portion of the breakwater. Further probes during the following seasons have verified the results of the first one and have supported the conclusion about the sophistication of Herod's harbour engineers.

It is apparent that the base for the 200 foot wide breakwater was built in large hollow compartments, leaving as much as three-quarters of the total volume of the part below the water empty and confined to about the surface level of the sea. As Josephus described it:

> "The impetuous south winds that beat upon them (the

Figure 4: An artist's rendering of the reconstruction of wooden frames used in building the breakwater of the Herodian harbour in Caesarea (Robert Teringo).

coastal features) rolling the sands that come from the sea against the shore ..." (Jewish Antiquity, XV, 336).

Knowledge of the characteristics of the coastal processes along this shoreline seems to have helped the ancient engineers discover a technique so far unknown elsewhere in harbour engineering. The hollowed compartments would have been left for a period of time long enough for the surge to fill them with sand to about sea level. Only then would the builders cover the entire structure with a layer of rubble upon which the upper structures would be built. It is not known how long they had to wait, but it could not have been more than 3-4 years, because we do know that the entire project of building the city of Caesarea and its harbour took only 12 years from the start to the inaguration ceremonies.

Figure 5: Digging through the embedded sand within one of the hollow compartments at the main breakwater of Caesarea. (Photo: M. Little).

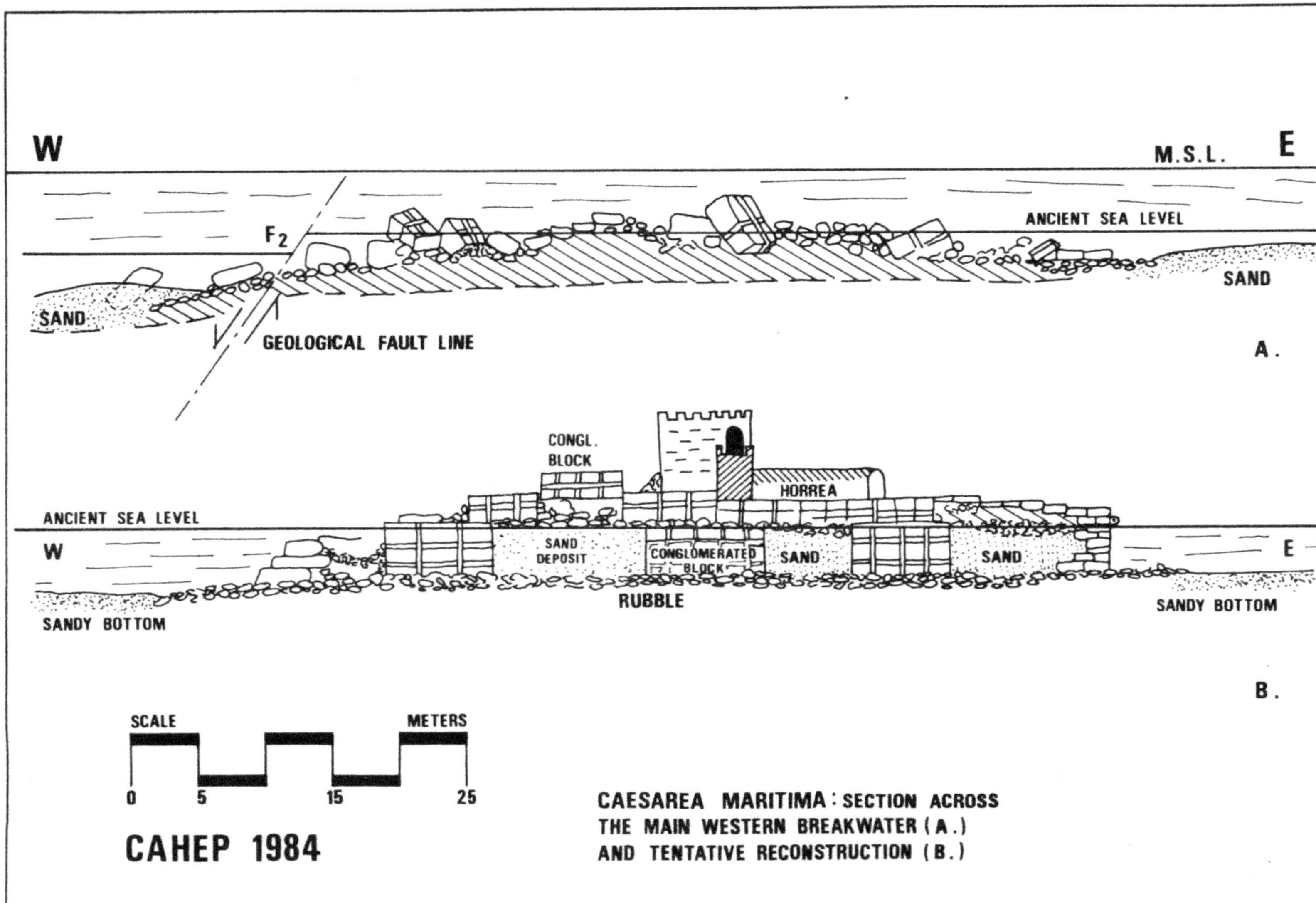

Figure 6: A tentative cross section of the main breakwater at Caesarea.

Within the laminated section of the sand-fill there are between five to eight interlayers of coarse shingled sand which may represent deposition of heavy winter storms. Taking the average annual number of such storms today would give us an estimated period of two to three years for such figures. (Figs. 6, 7).

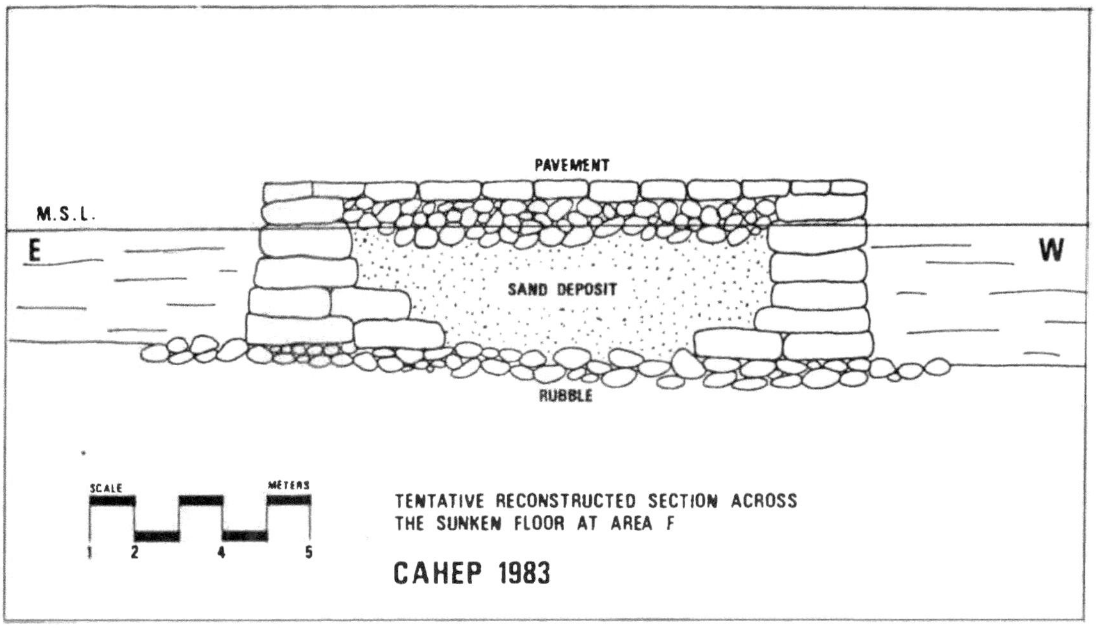

Figure 7: A cross section through a secondary jetty within the great Herodian harbour of Caesarea.

IV. VERTICAL SEA WALLS OF ASHLAR HEADERS

The above three technological features were associated with marine structures built on sand. The other environmental problem that the ancient harbour engineers had to deal with was the energy of the waves. It seems as if the most destructive effect of the waves upon marine structures was the vacuum effect of the retreating water from a vertical wall. Building such a wall from many relatively small blocks would risk its durability. Once a single block from the vertical face of the wall was sucked off by the surge it would only be a matter of a few years before the entire structure would be hollowed from the less consolidated fill beyond its façade. The opening would become wider and wider - until it would totally collapse. It seems that this destructive phenomenon was well-known since the earliest stages of the history of marine construction. The logical technological answer to this challenge would have been to build the sea walls in such a manner that each individual component would have maximum physical integration within the structure. This goal was met by the innovation of special types of ashlars: slim, long, and carefully-shaped stone blocks to be used in building sea walls. These blocks had to be laid in tight courses with their joints in a very close fit. These joints would be long enough to give maximum drag between any neighbouring blocks in the wall. In other words, the ratio between the face of each building

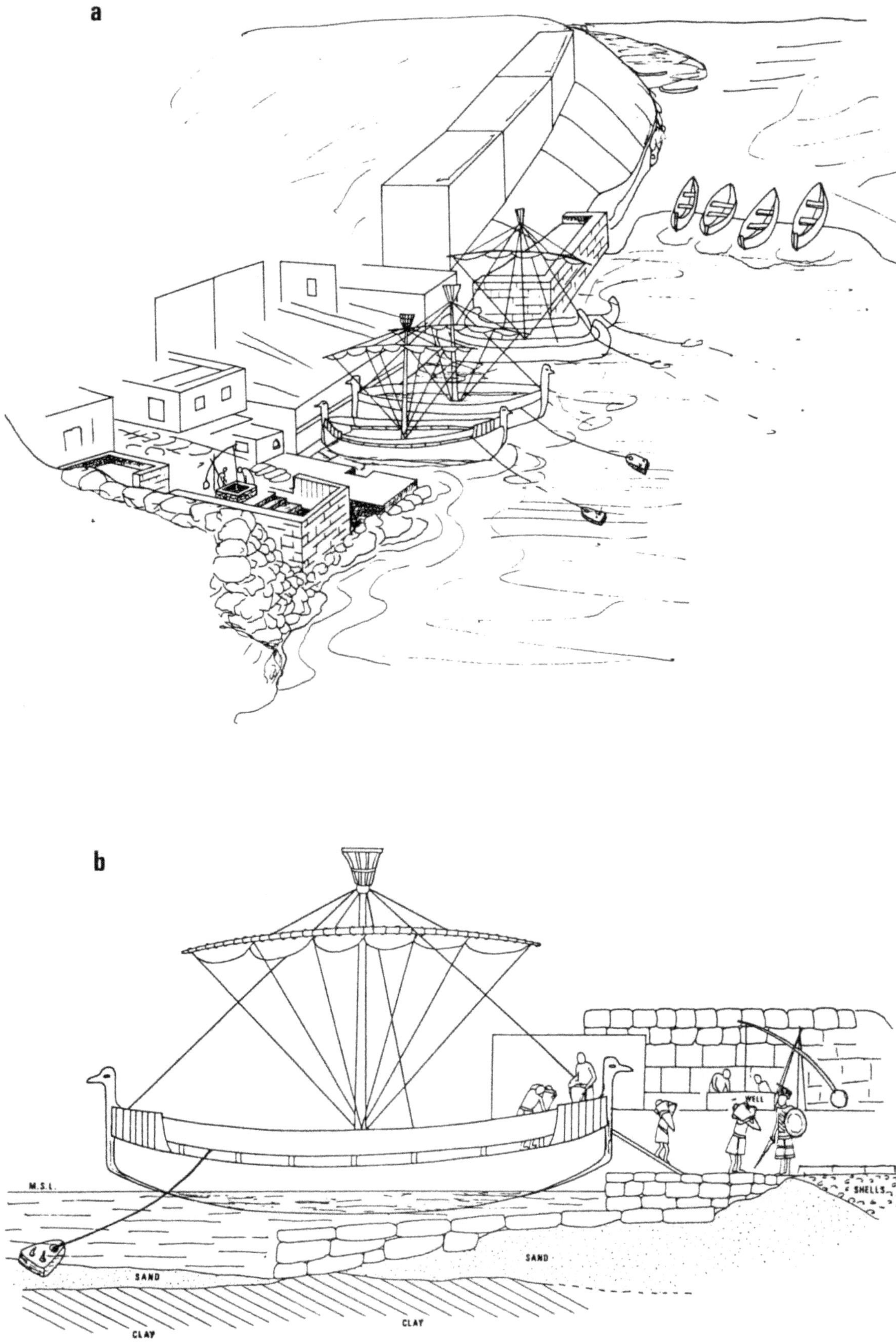

Figure 8: An artist's rendering of the Sea Peoples' harbour at Dor. (Z. Friedman)

stone and that part of it which was sunk into the structure would be as great as possible.

Till recently the oldest marine structure of this type was known to be the Phoenician mole which Braidwood excavated on the Syrian coast of the Mediterranean, at Tabat el Hammam. This mole, which was built very probably as early as the ninth century B.C.E., had facing edges of long, slim headers of 0.4 x 0.5 m and almost two metres of tight, smoothly cut joints(11). So, for every 0.2 sq.m. of the vertical face of the wall each block would have 3.6 sq.m. of tight joints. A similar ratio is to be found in the Phoenician moles of Athlit and Akko(12).

Though these moles are the earliest known so far to be built in this way, the very idea of building a "sea-wall" through the use of ashlar headers is to be found even earlier. Such are the foundation courses of the flanking buildings on both sides of the 13th century B.C.E. quay or paved landing stage on the south side of Tel Dor(13). This might be connected with one group of the Sea Peoples, the Sikuli, who operated this site as an international harbour, as we know from an Egyptian document of the following century(14) (see Fig. 8).

Yet the innovation of "Headers" in stone-made sea-walls was probably known at Dor much earlier. A massive wall of roughly cut stone blocks of considerable size (0.6 x 0.7 x 1.3m, on average) was found on the shores of the bay on the north-west side of Tel Dor. It seems to have been built originally as a base for retaining the man-made structures and to protect them from the sea-waves. The blocks of this wall were laid on the ground course with their shorter part facing the sea. Archaeologically they are dated to the earliest city on the site - to the 20-19th century B.C.E. (= MBIIa)(15) (see Figs. 9-10).

Surveying the locations which the Phoenicians had chosen for their vertical moles of ashlar headers, one can see that in all the sites (Tyre, Akko, Athlit) they were somewhat protected from the full surge of the south-western winter storms. In Athlit these moles are in the lee of a shallow rocky reef that causes breakage of great stormy waves away from the moles which are facing north and east, away from the inclement weather(16). At Akko and at Tyre, the moles face south, but are well protected by long shore reefs further into the sea(17.) These protected locations seem to have been sufficient for keeping the moles intact through the annual period of storms and enabled seaborne transshipment activities along both sides during a good part of the sailing season. The use of ashlar header sea-walls therefore promoted a unique harbour feature - a mole that was to be used as a jetty for quick and easy trans-shipment of goods.

V. WAVE CATCHERS

The fact that the early dwellers of port cities were aware of the disturbing and destructive effects of the storm breakers is no surprise. It seems that one of the main problems was caused by the

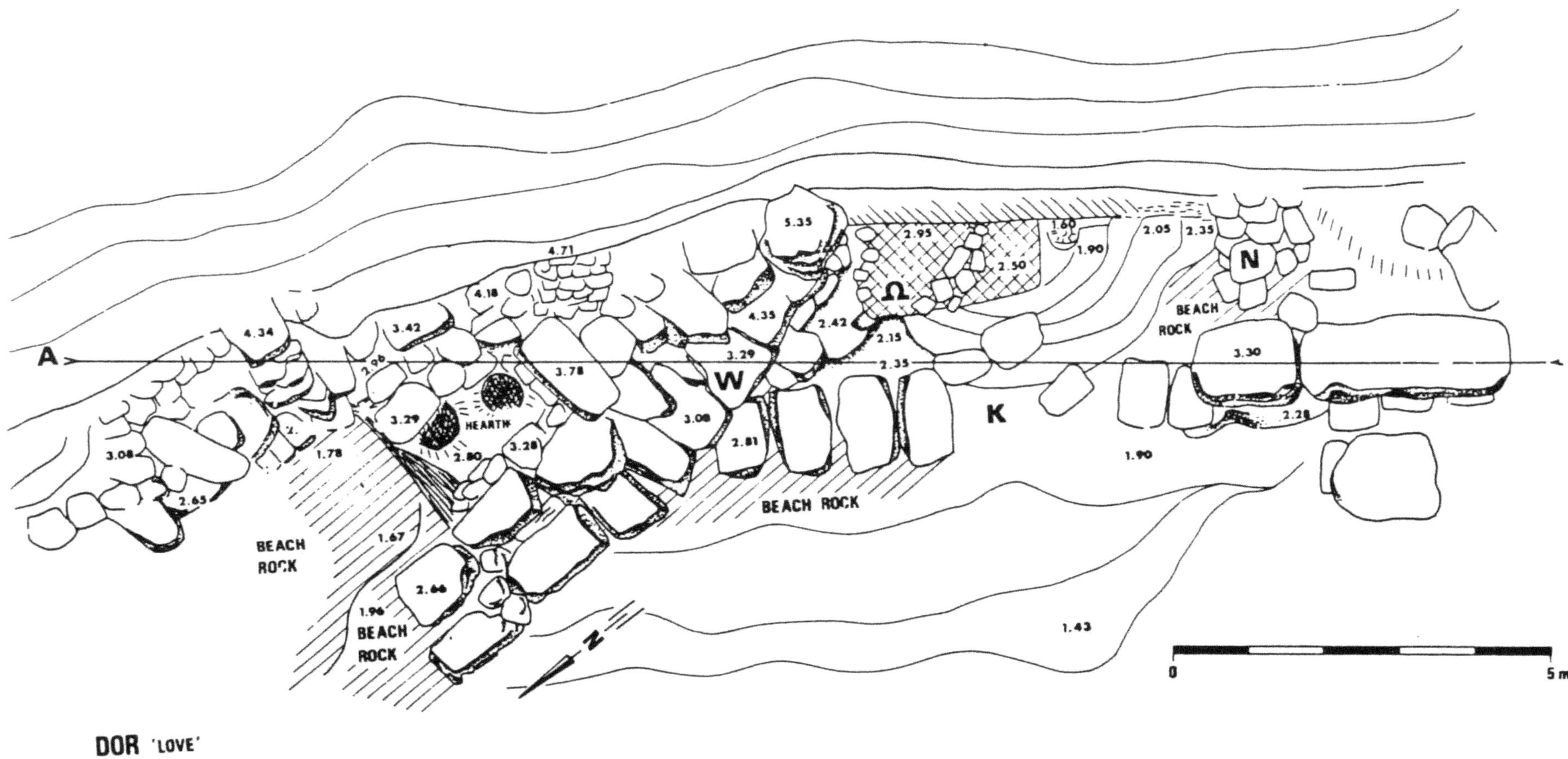

Figure 9: A plan of the MBII ashlar sea wall at Dor.

splashing of the salty droplets whenever the waves would break against the coastal cliff. The Phoenicians, who chose rocky headlands and promontories for their marine stations, were particularly bothered by such phenomena. So, in order to avoid this problem, and to enable their buildings to be located close enough to the water, they developed ways of solving, or at least minimizing the problem. The idea was to create hollows in the rocky shore, parallel to the water line, leaving the natural outcrop as a screen against the waves.

Such hollows would be used as an immediate source of building stone, yet would also serve as wave catchers that would absorb the sea water driven onshore by the breakers. The wave catchers would channel and drain the water off, back to the sea, leaving the lee side of the hollow dry. These devices are prominent features at Phoenician maritime installations such as the outer harbour of Sidon on the offshore island(18), and at each headland along the seafront of Dor (see Figs. 11,12)(19). This Phoenician innovation might have been relevant and of some significance in places where the natural topography of the port's location favoured the use of rocky headlands, or in coastal features which face the direct force of the marine elements. So, while planning a full-scale harbour complex in open waters, this concept could have been adopted in a special way.

VI. SUBSIDIARY BREAKWATER

A unique feature of the Herodian harbour at Caesarea, which seems to be an advanced application of the Phoenician wave catchers is the subsidiary breakwater. This feature can still be detected more or less parallel to the external side of the southern part of the main breakwater, some 25-30m into the sea. (see Fig. 13). At the present stage of the underwater research it is hard to tell whether it was an independent structure or an external incorporated part of the main breakwater. In either case it was obviously bounding a hollow space in its lee to exhaust the wave's energy before reaching the spinal wall. Being non-continuous, such a structure would provide an exit for the rip current and would prevent the surge from piling up towards the main breakwater. It is quite probable that this unique feature is what Josephus was referring to by using the term Procumatia which is not to be found in any other ancient Greek text(20). The subsidiary breakwater at Caesarea was built as a relatively modest structure. The width at its base was some 5-6m and only about 3m at what seems to have been its top, at about the ancient sea-level (e.g. about 5m below the present level).

The subsidiary breakwater in Caesarea was built upon a rubble-paved sandy bottom at what was then a depth of about 3m below M.S.L. It consisted of sandstone blocks, large masses of concrete, and a fill of rubble. It was similar to the main core of the great breakwater that was used as a partition wall between the hollow compartments sand-filled by the surge (see above).

Figure 10: A sea wall of ashlar headers at the landing stage of the "Sea Peoples" at Dor.

Figure 11: A wave-catcher at Dor.

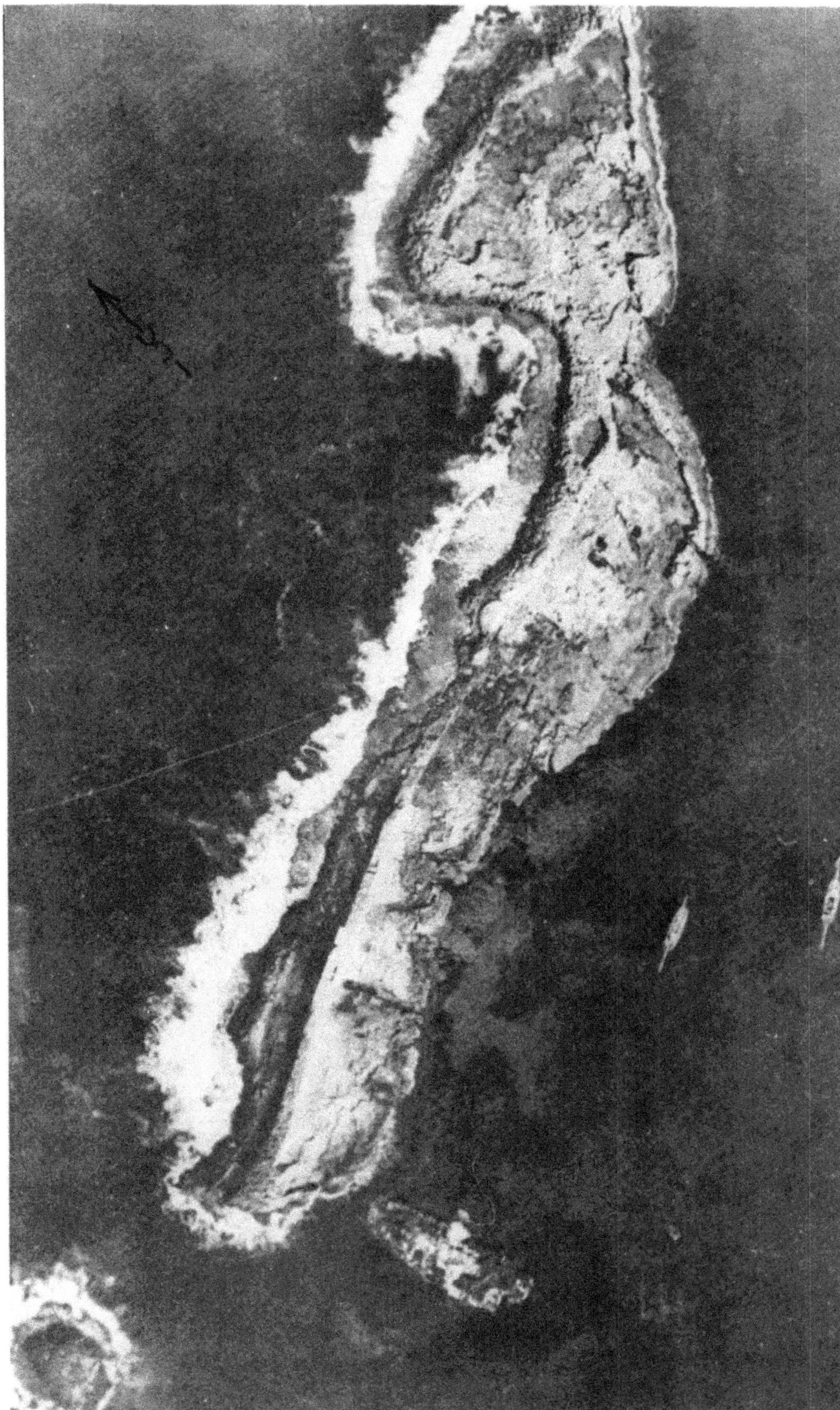

Figure 12: A wave-catcher on the off-shore harbour of Sidon.

VII. KEEPING THE HARBOURS SILTFREE

The instances of harbours having been so heavily silted as to be totally useless have been well-known around the relatively tideless Mediterranean all through the ages. Ephesos and Leptis Magna are but two of the better-known cases. Even today the problem of close basins being rapidly silted-up is a continuous problem for many commercial ports. In most harbours with tideless waters a permanent dredging is quite a common procedure. Yet such mechanical dredging would not have been possible or practical in ancient times. The only available alternative therefore would have been either to avoid silting, or to flush the harbour's basin in order to create a permanent outflow current that would clear off the silt. The first alternative would not be practical whenever the silting of the harbour is caused by long-shore, wave-carried sand. This is due to the simple fact that each wave, while deffracted at the harbour's entrance, would release the solid particles to be deposited on the floor of the closed basin.

A barrier up the coast, either natural or artificial, might slow down this coastal process but would not stop it in the long run. While in the case of riverine silt there are possible solutions, such as the alteration of the flow of the river, the damming of the river course(21). Yet, there is no practical way to avoid completely the gradual silting of a closed harbour along a sandy shore. This phenomenon seems to have been known to the ancient mariners and was to be encountered from the time artificially closed harbours were first built. The idea and the need for a closed harbour along the open shoreline seemed to coincide with the introduction of a competitive sea-borne trade and naval warfare. First the Phoenicians and then the Greeks were the initiators of these practices. They came of age during the Hellenistic era and had reached their heyday during the first and second centuries C.E. with the maritime endeavours of the Roman Empire(22). While in cases where the actual harbour was located in the lee of relatively shallow rocky reefs, the practice was of placing

Figure 13: A tentative cross-section of the southern breakwater of Caesarea.

detached breakwaters, with free passage for cross-reef wave energy to perpetuate flushing current through and out of the protected basin (as it was the case in Tyre, Athlit and probably Akko(23)), in the cases of less favourable locations or where a true "Limen Kleistos" was in demand, a more sophisticated device was to be used. The concept was to create a continuous volume of outflowing current through the entrance of the harbour by adding siltfree water to the closed basin over the breakwaters. These through-flushing channels would be controlled by a series of sluice gates, in order to meet the changing (though limited) tide gauge and wave energy. The flushing channels would be open to the sea with their base just above MSL, at the side where the sea bottom is rocky and shallow enough for the surge to break and release most of its sandy substance before reaching the breakwater. This is so in the case of the rock-cut channels across the levelled promontory on the west side of the closed harbours at Sidon(24), Dor(25) and Caesarea.

In Caesarea, the destruction of the southern breakwater, due to its considerable submergence and the later uses of its stable part, east of the fault line, left few fragmented remnants of its original surface. Yet, the opening of a flushing channel just above the present MSL can still be clearly seen as a rock-cut feature at the south-west corner of the natural promontory that was the stem for the main breakwater(26). At present the rest of the channel, its course across the promontory and its opening into the closed basin are covered by a modern cemented road. During the study of the ancient harbour, data

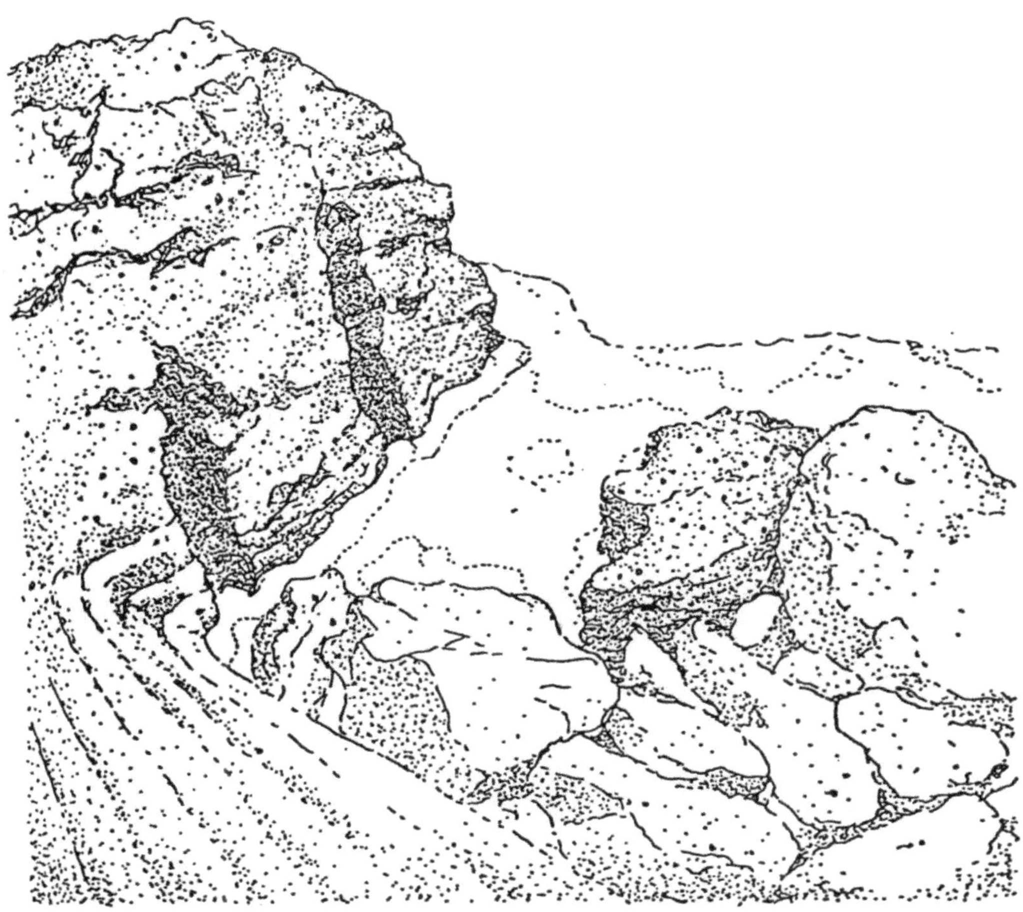

Figure 14: The sluiced flushing channel in Caesarea. (L. Vann).

were collected which verified that the successful flushing was active and effective (see Fig. 14). Whenever the probing excavations within the harbour's basin had reached a depth of 8.1 - 8.4m below MSL a well-defined compact layer of clay was exposed. Above it, there was sand and rubble and below - coarser sand with no pottery sherds. Whenever there were sherds within or directly on the clay, their date would be of the Herodian period, or of the following generation. It is quite obvious that such a layer (see Fig. 15) represented a depositional environment with minimum turbulence or wave energy. This layer and the pottery found within its context represented the sedimentation in the harbour basin during the time its breakwaters were still intact. The next data were exposed just outside the entrance to the harbour. There, a layer of about 2m was found, below the sand. This layer thinned off further away inside and outside of the narrow passage(27). Within this layer of mixed shingle, sand and mud was found; there was a dump of endless clay vessels, bronze pieces, wooden fragments and other typical garbage of an operational harbour. It can be deduced that this dump represented the outcome of an outflowing current which not only prevented the sand from coming into the harbour through its entrance, but even kept the floor of the closed basin free of jettisoned garbage(28).

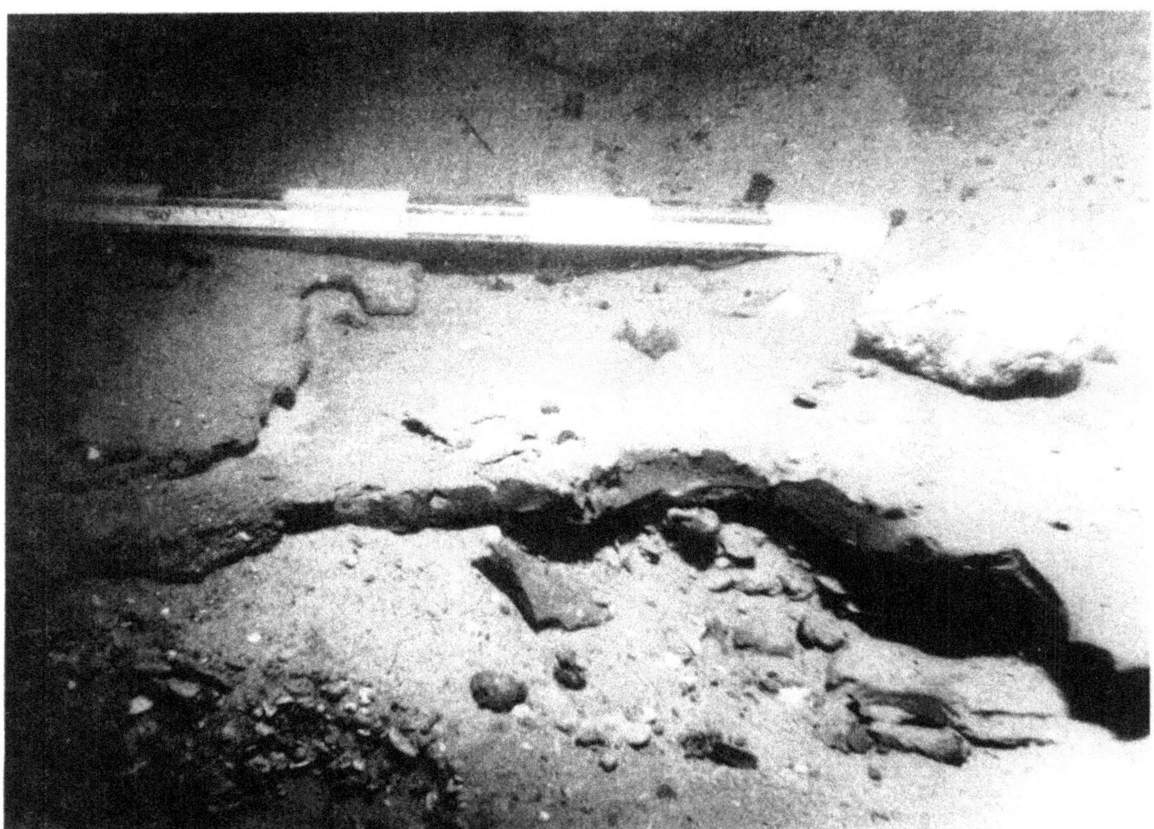

Figure 15: The muddy layer on the floor of the Herodian harbour in Caesarea. (Photo: H. Wadsworth).

The delineation of the artificial features just inside the entrance indicated that there was a funnel-like topography which would channel

Figure 16: An aerial view of the entrance to the Herodian harbour of Caesarea with its funnel-like feature. (Photo: R.L. Hohlfelder).

the outflowing current to fill the entire cross-section of the opening, leaving no free space for an inflowing countercurrent (Fig. 16).

VIII. WINTERING HARBOUR

The limitation of seasonal sailing all through antiquity might have caused a practical and economical problem. A metropolis dependent on sea-borne importation would suffer for a long period of being cut off from its current supply. In the case of Rome we know that the prices of imported goods would rise steeply during the winter. The first merchantman to come into one of the Italian ports from the Levant, bringing a fresh supply of goods from as far as the Indian Ocean would get an especially high price for commodities which were in great demand, such as spices, salt and aromatic plants. For this reason, a port in the Levant which might accommodate a considerable fleet of merchantmen over the winter, furnished with facilities necessary for loading them during the idle months of the winter might be favoured by the sea-borne traders. It would have enabled them to set sail towards Rome, already laden at the very beginning of the sailing season.

This fact might explain the sophistication of the Herodian harbour at Caesarea and the testimony to its being large enough to accommodate over 100 merchantmen along its quays and to them being loaded during the idle winter season. In order to facilitate wintering within a harbour basin, while being loaded with goods for export, Herod's engineers used their understanding of coastal processes in order to establish a quay free of salt water spray. At the Herodian harbour of Caesarea the subsidiary breakwater and the wave catchers would permit trans-shipment of goods to the mooring merchantmen through the whole winter period. The length and breadth of the main breakwater might have been designed to meet the demand for oriental and even tropical commodities which would reach Caesarea over the landbridge from the Red Sea. Having a breakwater broad enough to accommodate vaulted storage places would make the trans-shipment of transit goods in the harbour of Caesarea easier and faster than elsewhere in the Levant. By constructing a breakwater of multipurpose functions the adjacent city would be kept free of the problem of having a large transit trade carried through its streets. The harbour could therefore be operated as an autonomous unit, being active all through the year.(Fig. 17).

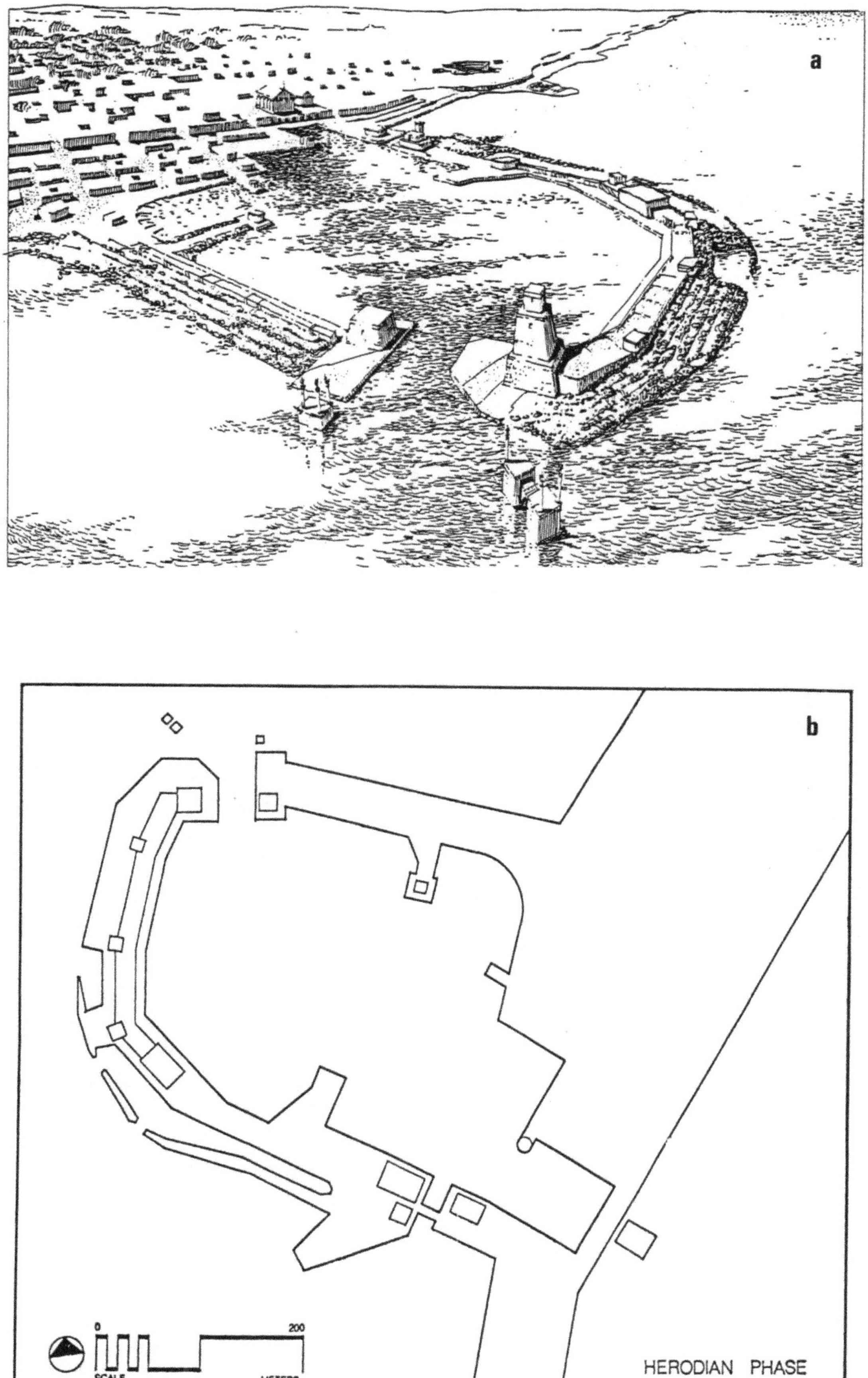

Figure 17: An artist's rendering the Herodian harbour and its plan (Steve Gianeti).

NOTES

1. The two recent scientific reports about the archaeological research at the Harbour of Caesarea are:
J.P. Oleson, R.L. Hohlfelder, A. Raban and R.L. Vann, "CAHEP Preliminary Report on the 1980-83 Seasons", Journal of Field Archaeology, 11 (1984): 281-305, and an extensive bibliography of other publications.
A. Raban, "Caesarea Maritima 1983-1984", International Journal of Nautical Archaeology 14.2 (1985): 155-177.
See also, A. Raban, Guide to Sebastos: The Ancient Harbours of Caesarea Maritima, Tel Aviv, 1982; R.L. Hohlfelder, "Herod the Great's City on the Sea: Caesarea Maritima", National Geographic Magazine, 171.2 (1987): 260-279.

2. The Caesarea Ancient Harbours Excavation Project (CAHEP) was initiated and conducted by the Center for Maritime Studies (CMS) at the University of Haifa jointly with the Universities of Colorado, at Boulder; Victoria, British Columbia (till 1985); Maryland and Rutgers, at Newark (since 1986). The author serves as the head of the project, with Prof. R.L. Hohlfelder, Prof. J.P. Oleson (till 1985), Dr. R.L. Vann, Prof. R.R. Stieglitz (since 1986) as co-directors.

3. See e.g. A. Raban, "The Ancient Harbours of Israel in Biblical Times, in idem (ed.) Harbour Archaeology, B.A.R. International Series 257, Oxford, 1985: 30-40.

4. The earliest dated Phoenician sea wall of this type was found at Tabat el Hamam in Syria; see R.J. Braidwood, Syria,21 (1940): 183-226.

5. A. Raban, International Journal of Nautical Archaeology, 12 (1983): 229-249; idem, "The Harbour of the Sea Peoples at Dor", Biblical Archaeologist 50.2 (1987): 118-126.

6. The case of the wooden caissons at Caesarea and the reference of Vitruvius are well dealt with by J.P. Oleson, "Herod and Vitruvius: Preliminary Thoughts on Harbour Engineering at Sebastos, the Harbour of Caesarea Maritima, in A. Raban (ed.) Harbour Archaeology, 1985: 165-172. For an illustrated reconstruction of CAHEP's see National Geographic Magazine, 171.2 (1987): 264-5.

7. See F. Grange, Vitruvius, De Architectura, Cambridge, 1931, 5.12.5.

8. See A. Raban (supra, note 1): 155-158.

9. See D. Neev (et al.), "The Young (Post Lower Pliocene) Geological History of Caesarea Structure", Israel Journal of Earth Sciences, 27 (1978): 43-64.

10. A. Raban, 1985 (supra, note 1): 161-163.

11. R.J. Braidwood, 1940, (supra note 4): 183-226.

12. A. Raban, 1987 (supra, note 3): 30-40.

13. A. Raban, 1987 (supra, note 5).

14. See "The Story of Wen Amun" in J.B. Pritchard, *ANET(2)*, Princeton, 1955: 23-25.

15. A. Raban & E. Galili, in *International Journal of Nautical Archaeology*, 14.4 (1985): 334-339.

16. A. Raban, 1985 (supra, note 3).

17. See D.J. Blackman, "Ancient Harbours in the Mediterranean, Part 1, *International Journal of Nautical Archaeology*, 11.2 (1982): 92.

18. H. Frost, "The Offshore Island Harbour at Sidon and other Phoenician Sites in the Light of New Dating Evidence, *International Journal of Nautical Archaeology*, 2 (1973): 76-79.

19. A. Raban, *International Journal of Nautical Archaeology*, 10.4 (1981): 287-308; "Some Archaeological Evidence for Ancient Maritime Activities at Dor", *Sefunim*, VI (1981): 15-26, The National Maritime Museum at Haifa, Israel.

20. See J.P. Oleson, "Herod and Vitruvius: Preliminary Thoughts on Harbour Engineering at Sebastos: the Harbour of Caesarea Maritima", in A. Raban (ed.) *Harbour Archaeology*, B.A.R., Oxford, 1985: 169-170.

21. And see A. Raban, "Artificially Alternated River Outlets along the Mediterranean Coast of Israel during the Bronze Age", in R. Paskoff and P. Trosset (eds.), *Déplacement des Lignes de Rivage en Méditerranée d'après les Données de l'Archéologie*. Aix-en-Provence, Marseille, in print.

22. See D. Blackman (supra, note 17): 186-193.

23. idem: 199, A. Raban, 1984 (supra, note 3): 35-38.

24. D. Blackman (supra, note 17): 199-202, fig. 9.

25. See Raban & Galili, 1985 (supra, note 15): 341-345.

26. A. Raban, Guide to Sebastos, 1982: 8; A. Raban & R.L. Hohlfelder, "The Ancient Harbours of Caesarea Maritima", Archaeology, 34.2 (1981): 59-60.

27. A. Raban, "Recent Maritime Archaeological Research in Israel", International Journal of Nautical Archaeology, 12.3 (1983):243-245.

28. A. Raban, 1985 (supra, note 1).

BRO. S. DOMINIC RUEGG

Minturnae: A Roman River Seaport on the Garigliano River, Italy

Minturnae was founded in 295 B.C.E. as a defensive garrison at the point where the Via Appia crossed the Garigliano River, 2 km. inland. There is some literary, inscriptional, and archaeological evidence that it was a river sea-port. The river is navigable today and has not changed its course since prehistoric times. Marius set sail from Minturnae to Africa in 88 B.C.E. 29 inscriptions from 1st century B.C.E. and others later in date were dedicated by merchants from Greece, the Middle East and Italy. A Republican one commemorates an architectus navalis and indicates that ships were built here. Archaeological artifacts from FG 1 demonstrate the importation of marble, lead, iron, copper and tin and local manufacture of iron artifacts. Exports consisted of Campanian black glaze ware and wine in Greco- Italic, Dressel 1, and Dressel 2-4 amphorae. FG 1 was probably the first area of the river used as a seaport and gradually diminished in importance as the harbour facilities were developed downstream and near the mouth.

INTRODUCTION

The purpose of this paper is to provide a preliminary examination of some of the literary, inscriptional and archaeological evidence which demonstrates that the little- known provincial city of Minturnae, Italy, was a Roman river sea-port on the Garigliano River (the ancient Liris). The term Roman river sea-ports refers to towns or cities which were located upstream some distance from the mouth of a river at which sea-going vessels loaded and unloaded their goods and passengers directly at river-side wharves (Cleere 1978). This does not exclude the use of barges for river traffic as was the general case on the Tiber between Ostia and Rome (Meneghini 1985; Mocchegiani 1982).

During Roman times there were several river sea-ports, as investigations have shown in England (Fryer 1973; Cleere 1978; Milne 1985), Northern Europe, Southern France, North Africa, etc. (Shaw 1972). In Italy, their existence seems rare, not because they did not exist, but because some rivers have changed course, often due to silting, probably caused by the rapid deforestation during the late Republican period (Vita-Finzi 1969, Walker 1969, Schmiedt 1975). For example, the Voltorno River was once navigable for some distance inland, and it is very likely that Capua had a port installation on the Voltorno as yet undiscovered. On the other hand, Pisa, some 10 km from the sea on the Arno River, was a Roman port and it continued to function into the 12th century C.E. (Pasquinucci and Rossetti 1986).

NAVIGABILITY OF THE GARIGLIANO/LIRIS

Was the Liris navigable and able to provide a safe harbour in Roman times? Is it in the same river bed as now? Or like many rivers in Italy has its course changed over the centuries? Today the Garigliano is navigable for about 10 km from its mouth and averages in depth from 7 to 10 metres. It is fed primarily by the streams Rapido, in ancient times the Gari, and the Liri, which rise in the mountains of the Abruzzi. Its sources include numerous springs and especially noteworthy are the voluminous springs at Lago di Posta Fibreno, which keep the river flowing strongly all year long and are sufficient to support several hydroelectric installations. Studies of the basin of the Garigliano River have clearly demonstrated that the geological formations above the modern auto bridge, upriver from the castrum of Minturnae, indicate that there have been few changes in the river bed from that point to the mouth since prehistoric times (Sevink, Remmelzwall, and Spaargaren 1984). In addition, there is no evidence of bradyseism along the beaches, but only of coastal erosion, such as at Sinuessa to the south and Sperlonga to the north.

The first historical reference to the navigability of the Garigliano is in 88 B.C.E. when various ancient writers note that Marius, fleeing from Sulla, fled to Minturnae by ship from Ostia and was then aided by a merchant of Minturnae, Beleus (Plutarch Marius 40, 1), whose ship was on the point of sailing to Africa (Bang 1910). In the time of Tiberius (14- 27 C.E.) the Digest xiii, 19, 2 (Bremer 1898) records a case which concerns a ship that capsized while trying to negotiate the mouth of the Liris. After the Saracens were defeated in 914 C.E. two defensive towers were built on the Garigliano, one near the mouth and the other 2 km upstream, no doubt to defend the river against further depredations by the Saracen pirates. In later times, the Liris was still navigable as far as Suio, six km further upstream from Minturnae, for the chronicle of Monte Cassino states that marble was brought by sea from Ostia to Suio for the building of the monastery in 1068 C.E. (Johnson 1933). In 1503, Piero de Medici drowned in the Liris while he was trying to raft four pieces of captured Spanish artillery downstream. He was a mercenary of the French who were defeated by the Spanish at the Battle of the Garigliano and is buried at Monte Cassino (Young 1923). In 1787 Cavaliere Domenico Venuti, a noted antiquities dealer, loaded a barge with antiquities. When the King of Naples discovered this, he demanded first choice; but Venuti claimed his barge had been lost at sea (CIL X 6012). In 1805-6 during the Napoleonic Wars there was some action on the Garigliano River when it formed a defence line against Napoleon during October and November 1805. At this time, the British had built a pontoon bridge across the river after Napoleon's overwhelming victory at Austerlitz. In June-July of 1806, a small squadron of British gun-vessels blockaded a French flotilla that had holed up in the Garigliano for about a month (Acton 1957). Recently, the Garigliano has been the home port of a few fishing vessels, but now has become a well-used harbour for motor boats and yachts, as it is still one of the few tolerable havens for ships along a coast singularly barren of shipping facilities.

HISTORY OF MINTURNAE

Minturnae is on the Via Appia about 150 km south of Rome (Salmon 1967, 1970; Constantini 1933). It was founded along with Sinuessa in 295 B.C.E. during the Third Samnite War 298-290 B.C.E. to prevent the Samnites from using the coastal route, either by land or sea, in order to join their allies in Umbria. Minturnae was the fourth citizen colony, known as a colonia maritima sent out by Rome. Citizen colonies were not primarily agricultural communities, as each citizen was not given enough land for subsistence farming, but rather they were strategic garrisons to guard the coast, sometimes for offensive purposes but especially for defence, as was the case with Minturnae. Whether or not there was a small contingent of naval forces at this time is unknown, but probably not. Besides being a coastguard station, Minturnae was strategically located on the Via Appia, which had reached the Liris River as early as 312 B.C.E. The Via Appia is about 2 km upstream from the mouth of the river and here the castrum, or military camp, still unexcavated, with the Via Appia as its decumanus, was founded to protect the river crossing (Frier 1969). A wooden bridge was built here some time around 250-200 B.C.E. (Ruegg 1983) and the river area just downstream from the site of the bridge was the original port of Minturnae.

Once peace was established in southern Italy, the history of Minturnae from the 2nd century B.C.E. to its demise in the 6th century C.E. was much the same as other ordinary Italian provincial towns. It was rather commercial in character, and an important stopping station on the Via Appia. It very probably never became a resort town with rich seaside villas like some towns just to the north, such as Formia, Gaeta, and Sperlonga. Quite likely it was a very minor rival to the main ports of Ostia and Puteoli. Minturnae shared in the general prosperity of the 2nd century B.C.E., and if the thousands of bronze coins, mainly minted in Rome, are any indication, this prosperity continued through the 4th century and even during the barbarian invasions which began in the 5th century C.E., despite a drop due to the disturbances of the 3rd century C.E.. The increase in the money supply, particularly bronze coinage, which was commonly employed for market exchange, suggests a healthy local trade (Renfrew 1984) and very possibly a somewhat extensive trade throughout the western Mediterranean.

In the 2nd century B.C.E. as the latifundia grew in the rich agricultural valley surrounding the Garigliano River, the wealthy landowners generally had their residence in town and Cicero mentions a few of them in his letters (Johnson 1933). They directed the political, religious, and social life of the town; erected imposing town buildings such as the fora, the temples, the theatre and the baths (de'Spagnolis 1981); and they worked their large landholdings with slaves (Tchernia 1971, 1981 and 1983; Frederickson 1981; Pleket 1983). Orosius (Histories V,9), relates that 450 slaves were crucified at Minturnae during the revolt of 135 B.C.E. The economy was slave-based and as in most of Italy, aside from Rome, farm produce and articles manufactured by urban craftsmen, were mainly for local consumption. Nevertheless, there is sufficient archaeological evidence in the river to indicate what some of the local manufactures were, the raw or refined materials that had to be imported, and the products that were exported (Hopkins 1983).

INSCRIPTIONS

There are several inscriptions which present some general considerations with regard to Minturnae as a river seaport. Two concern merchants, "mercatores". One is a dedication to the Emperor Nerva (96-98 C.E.) presented by some merchants at Minturnae from Beirut, Heliopolis, Ephesus, Tyre and Syria (CIL X 797). In the second, a Christian "puteolanus" merchant connected to and probably from Minturnae commemorates his son's death (CIL X 803). There is a tombstone in Bonn which can be interpreted as honoring a Roman as "conductor portus Lirensis", that is, the one who had the contract to collect import duties at Minturnae (Johnson 1933), although this is disputed. L. Burbuleius, a member of an illustrious family of Minturnae, discharged his duties in three other ports as "curator rei publicae" (CIL X 6006) and perhaps may have received his first training at his home-town port (D'Arms 1981; cf. Cavuoto 1982).

During 1931-33 when Dr. Jotham Johnson was directing the land excavation in the town centre, he uncovered in the foundations of Temple A, 29 inscriptions from the Republican era (Johnson 1933; de Rossi 1980) dedicated to Rome and Augustus, very similar in function to the ones found at Capua (Flambard 1983). Both give us the names of slaves and freedmen who are clearly linked to the names of the great proprietors known in these two regions. The inscriptions found at Minturnae mention two specific collegia or associations and seem to be cultic inscriptions of Greek and Middle Eastern merchants, who frequented the port of Minturnae in the last two centuries of the Republic. According to Johnson, the inscriptions are of families from the first half of the 1st century B.C.E.; According to Staedler (1942) they are from the resettling of the soldiers of Augustus in 28 B.C.E. at Minturnae (Lachman 1848). Nevertheless, there are a number of Greek names, such as Aphrodisius, Antiochus, etc. and five references to the socii picarii and five to the socii salinatores. The agents of these associations were freedmen or slaves, and the collegia were branch operations of the mother company in Rome, as they involved state controlled commodities. Most likely they were financed by the wealthy local urban and agricultural elite. With regard to the dealers in salt, this would seem to imply that the salt gathered at Minturnae was a commodity for export besides being for local consumption. Other dealers supplied pitch for ship-building, the sealing of wine jars, and for medical purposes. If pitch was available in quantity, then the pine forests in the mountains surrounding Minturnae were still in existence and the dealers in pitch probably also dealt in lumber, a very valuable commodity as Italy was largely deforested even in the 1st century B.C.E. (Johnson 1933).

A document, previously mentioned, concerns the navicularii who were transport entrepreneurs, owners of ships, or ship's masters. In a court case it was asked whether a navicularius who had charge of a cargo to be delivered at Minturnae, but who had to transfer the cargo to a second *navicularius* in another ship because the first was unable to come up the river at Minturnae, was responsible when the second ship capsized at the mouth of the river. There is no mention made of *nautae*, who were bargemen, though there can be no doubt that barges plied the Liris. Tommasino (1942) reported that other documents show that corporations of *attori*, administrators of public property (cf.Burbuleius above), of *naviculares maritimi* and of *fabri navales*, ship's carpenters, existed at Minturnae. Probably other corporations, such as caulkers and rope-makers, were also present (Rougé 1981; Bottigelli 1942). This brings us to the important question as to whether there was ship building at Minturnae? Johnson (1933) reported that there was a row of 15 shipways along the right bank, occupying about 200 m of the waterfront of Minturnae. They were built of reticulate masonry dating from the middle of the 1st century B.C.E. to the mid-1st century C.E. Certainly the protected waters of a deep river would have provided an excellent site for such ship building. Unfortunately, these shipways could not be found in 1967 or subsequent seasons and were probably covered over when the modern levee road was built.

The most convincing argument is an inscription, noted in the CIL X 5371 (Tommasino 1942), of Quintus Caelius, an *architectus navalis*, a master-builder of ships. This inscription is in private hands near Minturnae and was found in that region, "near the Liris in a field near the town of Sts. Cosmas and Damian". What is interesting to note is that Caelius was married to Camidia, who was the daughter of Aphrodisius, a Greek, who had been the slave of Carisius, and Aphrodisius is mentioned on one of the 29 inscriptions found by Johnson in 1931-33 and which date to the end of the Republic.

FIUME GARIGLIANO AREA 1: FG 1.

The archaeological area upon which we wish to concentrate is known as FG 1 (FG = Fiume Garigliano). It extends 200 m downstream from the oak pilings that I believe to be the first pier of the bridge, the Pons Tirenus, mentioned in a letter of Cicero written in 44 B.C.E. (ad Att. xvi, 13). This area of FG 1 parallels the west wall of the castrum and is about 12 to 14 m wide. The entire area is covered with hundreds of sherds of broken pottery and in addition to thousands of Roman coins, there are many objects made of bronze, lead, and iron. The reason that FG 1 has not been swept clean of artifacts by the river (as the 40 m of the centre has been) is because there is much fallen Roman masonry, probably from the docks and buildings that once lined the river bank. In addition, there are many large rocks which were thrown in during the 1930's to shore up the modern levee road along the right bank. Due to these obstructions, the method of excavation was primarily surface collection (Ruegg 1983).

ARCHAEOLOGICAL ARTIFACTS OF FG 1.

More specific arguments from the archaeological finds of FG 1 assist in demonstrating that Minturnae was a river seaport. This will be a rather summary description of some of the archaeological artifacts found during the five seasons of underwater excavations between 1966-1981, with special reference to artifacts in stone, iron, lead, bronze and pottery and whether these items are related to local manufacture or overseas trade.

STONE

Typically, a great deal of marble was used in the embellishment of Minturnae, especially from the end of the 1st century B.C.E. Marble makes its first appearance in the <u>Lex Rufrena</u>, an inscription of c. 44 B.C.E. (Cavuoto 1982). A great deal of rather ordinary statuary has been recovered from the land excavations, notably by Venuti in the late 1700's, Nugent in 1817, and the University of Pennsylvania expeditions of 1931-33. These can be viewed in the Antiquarium at Minturnae, the Museo Nazionale in Naples, the University Museum in Philadelphia, and the largest collection, in the Archaeological Museum in Zagreb. From the river excavation comes one statue of a headless togated male and nine statuettes (Harnett 1986), plus a number of marble fragments of columns, capitals, trim, flooring, etc. There still remain on the bottom some large marble pieces, for which there was not the equipment to raise, such as two statue bases, 21 large column drums, and two capitals. The predominant colour of this marble is white and presumably it came from further north in Italy, though it could also have come from some quarries in Turkey and elsewhere. There are only two examples of golden marble trim. The main point here is that there are no local sources of marble and all had to be imported by sea (Johnson 1935).

With regard to the 11 architectural pieces in tufa or limestone, the notable ones being a small altar and the other a composite capital of 2nd to 4th century C.E.,the case could be quite different from the marble. There are many tufa beds in the surrounding hills, but blocks would have had to be transported by land first and then barged down the river to Minturnae. Supplies of tufa could have been brought in more cheaply by sea, as was done even in modern times for the seaside development of Scauri, a nearby town, from the well-known quarries of the island of Procida near Naples (Johnson 1935).

IRON

Two pigs of iron were found in the river. More importantly, there were some pieces of haematite, a very high grade iron ore. This was probably imported from Elba during Republican times, or later from Spain (Forbes 1964; Tylcote 1976). There is quite a range of iron artifacts found in the Garigliano, such as spikes and nails; some of which are clenched (Ucelli 1940; Scrinari 1979; Joncheray 1970). One has a lead covering over the head and part of the shaft and may have been used in shipbuilding. There are iron hooks, one of which appears to be very similar to a boat hook found in Belgium, dated 1st-3rd centuries C.E. (de Boe 1978), a spearhead, a cotter pin, needles, knives and even some jewelry. Since there is such a variety of iron artifacts, it might be asked how Minturnae became a provincial locale for iron manufacture? The earliest reference comes from the times of the Punic Wars when the city of "Populonia furnished iron (probably from the Elban mines) for the levy in 205 B.C.E. for Scipio's invasion of Africa. In the war period from 200 to 150 B.C.E., the manufacture of arms was probably quite extensive and took place in the towns between Rome and Capua (Forbes 1964), "of which Minturnae was one of the more prominent. There is a small pair of iron tongs, which once had wooden handles, and was used in iron working (Gaitzak 1980; Mingazzini 1933). In addition, the Vescian Plain around Minturnae was intensively cultivated and many iron implements and tools were in demand. Cato (234-149 B.C.E.) mentions in his De Agri Cultura 135, 1 (175-150 B.C.E.) nine kinds of tools that he could purchase at Minturnae, and nearby Cales for use in his olive orchard. Several tools, a pick, a mattox, and a hoe, have been found in the Garigliano and these could have been purchased locally or perhaps were exported to other places.

One of the more interesting iron finds are the shoes at the foot of the more than 150 oak piles, which very likely form the first pier of the Roman wooden bridge, first built about 250 B.C.E. and in use until its destruction sometime in the 6th century C.E. Similar iron shoes have been found on Roman bridges or riverine foundations in Yugoslavia, southern Germany, France and England (Cuppers 1969). The eight iron shoes retrieved from the Garigliano are of similar construction: four flat pieces of wrought iron, each about 40 cm long. The lower half was hammerwelded and shaped to a point; and the upper half of the four flat pieces was spread out and became the arms which were then attached by iron nails to the tapered ends of the oak pile (Campbell and Fahey 1984). The iron was most probably smelted and made into flat pieces of the required length at Minturnae.

This was also a very convenient way to ship prefabricated iron bars. There is a shipwreck, Bagaud 2, off southern France, 2nd-1st century B.C.E., which has as part of its cargo bars of iron of approximately the same length as the bars used to make the iron shoes found in the Garigliano (L'Hour and Long 1984; for other ships with cargoes of iron bars, Gibbons and Parker 1986 IJNA, 15).

LEAD

There was a large amount of lead found in the Garigliano River. This is not surprising in itself as lead was the most common metal used during Roman times. It was easily found and was imported by ship to Italy in the form of ingots from, among other places, Sicily, Spain, Great Britain and Sardinia. In general, the Romans exploited one area at a time and then abandoned the mines in a particular area when they became too costly to operate (Boulakia 1972). One interesting artifact is a lead or pewter brooch decorated with three fish which may be an import. The only similar design is on two pewter spoons in the Museum of London (Walbrook no. 19190 and Mol no. 84.453/2) and all three were probably manufactured in England. No pigs of lead were found in the Garigliano in contrast to the large quantities of scrap lead.

The surprising thing is the number of uses to which lead was put: votive statuettes, vessels, stoppers, handles, keys, vasefeet, sling bullets, etc. How one would use a lead needle or a chain made up of lead links, or lead nails (though someone suggested lead nails could be used as caulking) is still rather mysterious. There follows just a few examples of moulded lead objects that might have to do with shipping:

Tesserae or small lead tokens or seals, about the size of a coin, with some sort of inscription or illustration, were used as tickets of admission to the theatre, as ballots, etc.; but they have also been found at ports, such as Rome Beirut and Lyons (Rostowzew 1903; Grenier 1934). Hence they were used as custom seals or commercial stamps, especially those with the remains of arms. They serve as witnesses to seaborne commerce and as a means of possible identification of imported and exported merchandise.

Lead Rings: the smaller ones are probably sail rings, but three larger rings deserve closer attention. One has a diameter of 8.9 cm with a hole in the ring and weighing about 275 gr. It is similar to one found at Grand Congloué and was quite possibly used as a part of the rigging (Carre 1983). The second has its eye on the outside of the rim, a diameter of 12.1 cm and weighs 980 gr; and the third has no eye, a diameter of 11.5 cm, and weighs 2.630 kg. These larger and heavier lead rings, especially the third one, might possibly be a line or net "freer" rings (Laures 1985; Ruegg 1986).

Lead Sheathing: there are some pieces of lead plates with nail holes irregularly spaced. They may have been used as small patches on hulls.

There is only one moveable lead stock and three stone anchors. There are two hemispherically shaped weights with circular depressions in the base which could be sounding leads. All told there are 262 lead weights. Five of them are marked with the Roman numerals I, III, and V, and each weighs approximately that many librae, that is, 1, 3 and 5 Roman pounds. 185 of these lead weights are definitely fishing weights and fishing must have been an important industry at Minturnae. Votives of lead and bronze clamshells were tossed into the river for good fishing fortune or in honour of Poseidon and Venus.

BRONZE

In addition, about 140 bronze fish hooks of various sizes have been found, some rectangular in cross-section, others circular, and three, a combination of both. There are a few that do not have the usual tang at right angles to the shaft, but the ends of the shafts of five of the hooks are serrated. Nets were also used for fishing as there are 16 netting needles as well as lead net weights. Among the pottery are sherds of amphorae which would have been used to export salted fish and garum or fish-sauce.

There are 92 needles of rather heavy manufacture which were needles for making sails and one awl with a lead handle which could have been used to make holes through rather thick sail cloth (Carre 1983). It can be compared to the wood-handled iron-tipped awl found in the Porticello wreck (Owen 1971).

Items that may have been used on ships include bronze washers, rings, cotter pins, eye-bolts, and a large hook with a duck-shaped head. There are over 600 bronze tacks and nails, some of which are clenched. There are nine tacks, decorated on the underside of the head with 4, 6, 8, or 10 lines radiating from the shaft and in the sectors between the lines are one or two raised dots. These are typical tacks used for nailing lead sheathing to the hulls of ships, such as found at Nemi and Ostia as well as port sites in England and southern France (Ucelli 1940; Arthur 1976; Bushe-Fox 1928; Bouscaras 1974; Fougère 1981).

No complete steelyards for weighing commercial items were found in the Garigliano, but there were 62 parts of large and small ones. One is an interesting bronze fitting, sometimes wrongly labelled in catalogues as "the top of a standard" (Walters 1899, nos. 2909-10), but which was designéd to check the amplitude of the oscillations of a large steelyard. Smaller steelyards are more numerous: bars with 6 or 7 incised lines to mark the scale, hooks, chains and weights. Steelyards are commonly found at Roman commercial sites and wrecks.

All of the copper and tin needed to make bronze had to be imported, probably as ingots of bronze, but none such were found during the excavation.

POTTERY

(Peacock 1977, 1982). Of the fine wares there are hundreds of sherds of Campanian Black-Glaze Ware mainly from the 3rd and 2nd centuries B.C.E. Of the 118 sherds collected (we tried to limit ourselves to recognizable shapes), 65 (55%) were identifiable (Morel 1969, 1981; Gianfrotta and Pomey 1981). Many of the shapes are the same as those found at the c. 200 B.C.E. kiln-site in Minturnae (Lake 1934-35), are also similar to the shapes at Cosa (Taylor 1957), and those found in the wreck at Cala del Diavolo, Tuscany (Martelli 1982). Besides the common stamps of rosettes and wheels, there are a number of other stamps which have not as yet been identified. The examples of this ware found in the Garigliano are not mis-fired pieces, which would indicate dumpings from a kiln, but are well-formed and certainly this ware was one of the exports of Minturnae, as a secondary cargo to wine, during Republican times.

The other type of fine ware is terra sigillata of which only 23 examples were found and these date from the 1st to the 2nd century C.E. This is in a way surprising because one would have thought that as the exportation of Campanian Black Glaze ware slacked off toward the end of the 1st century B.C.E., the amounts of terra sigillata, imported locally in Italy and from abroad, would have increased more dramatically; but so far, not many examples of this ware have been found in excavated Roman wrecks (Gianfrotta and Pomey 1981). The few pieces from the Garigliano are Italian and Gallic with one possibly Samian. The finest find is half of a bowl, Dragendorff Form 37, which first appeared during the reign of Nero and was the predominant bowl in the 2nd century C.E. (cf. Oswald and Pryce 1920; Atkinson 1914; Oxé and Comfort 1968).

There is a very large selection of ordinary coarse ware, jugs, cooking pots, plates and bowls. Most of the 172 samples are difficult to date, but a few are similar to pieces of coarse ware from Minturnae (Lake 1934-35) and also from Cosa, which date from the 2nd century B.C.E. to the 1st century C.E. (Dyson 1976). In addition, there are many examples of similar pieces from the land excavation of Minturnae in the 1930's, now deposited in the Museo Nazionale at Naples. It is generally believed that coarse kitchen ware was not exported, but the late Republican wreck of La Madrague de Gien furnishes definite proof of their maritime transport for commercial purposes and the reexamination of wrecks with quantities of coarse ware, aside from amphorae, will probably make this clearer in time (Tchernia, Pomey, et al. 1978; Gianfrotta and Pomey 1981).

There are two types of coarse ware that may be of note. The first are the so-called pot-lids, of which some 22 examples were found in the river. The more conically shaped ones were also found at Congloué in connection with Italic amphorae (Benoit 1961); and the flat ones are like those found in the wreck of a ship coming from Spain, Port-Vendre II, dated 41-42 C.E. (Colls, et al. 1977). The second are the dolia. These enormous jars were used to stockpile raw materials on Roman farms as well as in the warehouses at the ports, but have also been found on shipwrecks (Mocchegiani 1986). These ships were the bulk carriers of their day. For example, the wreck Grand Ribaud D, had about 10 dolia which held 1,500 to 2,500 litres apiece, probably of new wine, plus a 100 or so wine amphorae, Dressel form 2-4, from Campania (Hesnard 1984). The wreck of Petit Congloué with amphorae from Spain and Gaul had in its centre on the keel three rows of five or 15 dolia. The walls of these dolia had been braced with large lead clamps, just like six clamps found in the Garigliano. This shipwreck is from the 1st century C.E. and the dolia are stamped with various names of the Pirani family, who ran a pottery factory in Minturnae. The name "Piranus" is also stamped on a dolium in another shipwreck La Garoupe in the Antibes, southern France (Fiori 1972) and also one found off Corsica (Gianfrotta and Pomey 1981). The family of Piranus is mentioned four times on the Republican inscriptions found by Johnson (Johnson 1933). Since dolia would not be unloaded like amphorae, it is possible that these ships originated at Minturnae.

Of the 228 remains of amphorae, 118 or 52% were identifiable and 37 of these have very clear parallels. There are very few whole amphorae. In spite of the fact that we have 29 cultic inscriptions of Greek and Middle Eastern merchants from the 1st century B.C.E., there are only two examples of Greek amphorae, both from Rhodes. One has a fine stamp of IMAS with a caduceus, and Imas was active c. 150 B.C.E. (Grace 1970).

The only other items from Greece are three fragments of Attic lamps from the 3rd to 1st centuries B.C.E.

The earliest non-Italian amphorae from the Western Mediterranean are three fragments of Mana C2 with its wide flaring lip. These are late Punic, abundantly documented in cemeteries at Carthage, and are dated anywhere from the 3rd to 1st centuries B.C.E. (Sienna 1977).

The vast majority of amphora fragments are Greco-Italic, 3rd to 2nd centuries B.C.E., and Dressel 1 A, B and C, widely diffused in the Republican era and no longer used in the Early Imperial period. These are well documented at Ostia (Panella 1970), and from shipwrecks and land sites in Italy, France, and Spain (Joncheray 1970; L'Hour and Long 1984; Tchernia 1983; Gianfrotta and Pomey 1981). They were used to transport wine; and of the famous wines from Campania, the nearest to Minturnae were Vinum Massicum and Vinum Falernum. A Dres. 1B, found in Spain, is marked FAL MAS, i.e. Falernum Massicum (CIL XV 4553) dated 102 B.C.E., which probably came from the area of Minturnae. On the stopper of a Dressel 1 B wine amphora found in the shipwreck Dramont A, dated the first half of the 1st century B.C.E. is the stamp of Sextus Arrius (Santamaria 1975), a member of one of the important families of Minturnae, mentioned three times on the Republican inscriptions (Johnson 1935). Another important family registered once on the inscriptions (Johnson 1935; Cicero ad Att. v, 3) is that of Cornelii Lentuli. The Lentuli stamp occurs on Dressel 1 B amphorae throughout the Mediterranean and gives evidence of a vast commerce in wine, which very probably originated in the area of Minturnae (Boucaras 1974; Gianfrotta 1982). For this same period there is one variant of a late Dressel 1 B, a type denoted as Pascual 1, 30 B.C.E. to 150 C.E., which is from Catalonia and was used to export wine from Spain (Beltrán Lloris 1978; Solier, et al. 1979).

The next principal grouping of amphorae is Dressel 2-3-4, with the typical rolled rim and double rolled handles of these forms, and may have been produced in imitation of the wine amphorae of Kos, just like those of this type produced at Pompei, stamped L. Eumachi (Panella and Fano 1977). They date from the 2nd century B.C.E. to the end of the 1st C.E. at a time when the Italian vineyards were competing with vineyards of France and Spain (Tchernia 1971; 1981). The analysis of the clay clearly demonstrates that they were made in Campania, and not Spain (Tchernia and Zevi 1972). A Dressel 2-4 amphora, manufactured in Pompei, is stamped obliquely on the foot MAROF (Panella and Fano 1977). This type generally contained Falernian wine. This wine amphora plus a few others of the same period from other parts of Campania, Apulia and Calabria (Baldacci 1972) imply that Minturnae was involved in transit trade.

After these dominant Campanian types of amphorae, Greco-Italic, Dressel 1 A, B and C and Dressel 2 to 4, all of which were used to export wine, there is a sharp drop in the number of sherds of later amphorae, both Italic and non-Italic. For example, there are three sherds, all of different types, but dating from before the end of the 2nd century C.E., which were made in Spain and used to transport garum or fish-sauce and other varieties of salted fish, as well as oil. There are also some examples from North Africa, such as Tripolitana I and II (Panella 1973) and Panella form LXI (Panella 1973). Two of these date from before the end of the 1st century C.E., one for the transport of garum and the other of preserved fruit; and two from before the end of the 3rd century C.E. After that, there are only two examples from the 3rd to 5th centuries C.E.

From this brief survey of the pottery of FG 1, the following four points can be made:

1. The Black Glaze ware is Campanian and mainly from the 3rd-2nd centuries B.C.E.

2. The terra sigillata is primarily Italian and Gallic from the 1st-2nd centuries C.E. and in rather small quantities.

3. The pottery, as to be expected, is largely coarseware and overwhelmingly composed of amphorae. The greatest period of activity was from the 2nd century B.C.E. to the 2nd C.E. These amphorae are, for the most part, Italic with a few coming from Spain and North Africa.

4. There is a surprising lack of late Roman amphorae in FG 1.

CONCLUSIONS

There are two conclusions that can be drawn from these four points:

First, the amount of Campanian Ware and of coarseware, mostly Italic amphorae, suggests that FG 1 was a mercantile area and not a domestic one; that some volume of trade was carried on from the 3rd century B.C.E. to the 2nd C.E. especially in the western Mediterranean; that Minturnae was a very Italian site. Even though it did import products, like wine and salted fish products from Spain and Gaul, and a bit later preserved fruits and garum from North Africa, Minturnae in general did not look far afield for its products.

Second, the fact that there are very few late Roman amphorae suggests either that Minturnae became a much less important seaport after the 1st-2nd centuries C.E., or that the harbour facilities were enlarged in some way. The second alternative is more probable for the following reasons:

1. Minturnae grew to be quite a large and prosperous provincial town during the late Republic and Early Empire. It was surrounded by a wall with 24 towers and spread to both sides of the Garigliano River (Costantini 1933; Lachmann 1948).

2. The course of the river from above the originally founded castrum to the mouth has not changed since prehistoric times, as had been clearly demonstrated by Sevink and his students (Sevink, Remmelzwaal and Spaargaren 1984).

3. Very probably, at some time during the 2nd century B.C.E., the Romans began to line the river banks with wharves. At FG 1 there are scanty remains of at least two in stone; but further down the river, on the left bank, are two sections of wharves. Unfortunately one, which had many amphora sherds at its foot, has been covered up by a recently constructed marina, but the other is about 30 m long and constructed in two different styles. The foundation of the downstream section, about the first 15 m, is made of opus reticulatum and surmounted by a brick wall about 50 cm. thick; the foundation of the upstream section is of concrete rubble construction and surmounted by a brick wall, about five bricks in height. Within the brick wall are rectangular insets for vertical oak posts which, in turn, hold up the topmost section of the whole retaining wall, a series of horizontal oak planks held together by mortices and tenons. A random sample of one of the oak posts was analyzed by C14 and the date is Roman: 30 B.C.E. +/- 60 years (R-1082a). About 100m upstream from the mouth on the right bank, just opposite the so-called Temple of Isis, which is now fenced in on private property, is a wharf of limestone blocks (cf. Mocchegiani: 1982) and in the river bank a retaining wall of opus reticulatum 1st century B.C.E. to 1st century C.E.

Downstream from this wall are four other retaining walls in the river bank, each one of a different construction. One of these is similar to the oak wall described above and again, a random sample was taken from an oak post and analyzed by C14, and dated as Roman 260 B.C.E. +/- 70 years (R- 1086a).

4. When a cement block factory was operating in the late 1960's at the mouth of the river, its dredge brought up many sherds of late Roman amphorae.

5. In 1966 when we were conducting a sonar survey of the river near the cement block factory, just inside the mouth, the strip chart showed that the bottom was reflectant and reverberant, indicating a mass of broken up objects. Unfortunately, there is a heavy layer of sand over this unidentified stratum, much too deep to investigate with the equipment that we had at our disposal. I suggest, that the sonar had picked up a stratum of broken amphorae, very probably of late Roman date.

6. Just at this point, immediately over the modern levee, was a fresh water lagoon, which was probably a part of the habour facilities. Unfortunately, it is now almost totally filled in and used to stockpile such items as prefabricated cement pipes. Old maps from the 17th and 18th centuries still show this lagoon connected to the river. In the story of Marius' escape from Sulla in 88 B.C.E., when his ship arrived at Minturnae, the sailors after casting anchor at the mouth of the

Liris where the river expands into a lake (lagoon?), advised Marius to leave the vessel (Plutarch 37, 3), not only because they were afraid of Sulla's wrath, but perhaps also to give Marius an opportunity to seek refuge in the Temple of Marica (Mingazzini 1933). Later, when he was allowed to escape, he was taken through the sacred grove of the goddess Marica to a ship; and this lagoon is just beyond her Temple. Cores taken in this lagoon in 1969 suggest the recovery of some remains of Roman ships (Remmelzwaal, 1978). Remmelzwaal claims that during Roman times, this lagoon was used as a harbour for Minturnae and that it possibly was partly artificially constructed, as were many small harbours during the 1st-2nd centuries C.E., like the inner basin at Marseille, Terracina and at nearby Monte Argento, (cf. Rougé: 1981). At present, 2 m or more of clay and peat cover any Roman remains in the lagoon. This lagoon has still to be investigated archaeologically.

In conclusion, I have tried to demonstrate from literary, inscriptional and archaeological evidence that Minturnae was a Roman river seaport and in particular that the area of FG 1, just below the bridge where the Via Appia would have crossed the Garigliano River, was the "old" or first seaport of Minturnae from the 3rd century B.C.E. and gradually diminished in importance during the early part of the Common Era as the harbour facilities expanded downstream.

REFERENCES

Acton, H. 1957. The Bourbons of Naples (1725-1825). London: Methuen & Co. Ltd.

Arthur, P. 1976. Some recent finds from Ostia. Spink's Numismatic Circular, March.

Atkinson, D. 1914. A hoard of Samian ware from Pompeii. The Journal of Roman Studies 4: 27-64.

Bang, M. 1910. Marius at Minturnae. Klio X: 178-191.

Beltrán Lloris, M. 1978. Las Anforas Romanas en Espana. Zaragoza: Institución Fernando el Católico.

Benoit, F. 1961. Fouilles sous-marines. L'Epave du Grand Congloué à Marseille. Gallia, suppl. XIV.

Bottigelli, M.D. 1942. Ricerche epigrafiche sulla marineria nell'Italia romana. Epigraphica 4: 69-87, 143-154.

Boucaras, A. 1974. Les marques sur amphores de Port la Nautique. Cahiers d'Archéologie Subaquatique III: 107-8.

Boulakia, J.D.C. 1972. Lead in the Roman world. American Journal of Archaeology 76: 129-144.

Bremer, F.P. 1898. Iurisprudentiae antehadrianae. Leipzig: Tuebner.

Bushe-Fox, J.P. 1932. Excavations in the Roman Port at Richborough, Kent. Oxford: Society of Antiquaries.

Campbell, J. and Fahey, F. 1984. A Metallurgical Study of Iron Shoes from the Piles of the Roman Bridge at Minturnae. The Journal of the Historical Metallurgy Society 18: 21-30.

Carre, M.B. 1983. Gréement et equipment des navires Greco- Romains d'après les document archéologiques I-II(doctoral dissertation). University of Aix-en-Provence.

Cavuoto, P. 1982. Iscrizioni latine di Minturno. Ottava Miscellanea Graeca e Romana 33: 489-585.

Cleere, H. 1978. Roman harbours in Britain south of Hadrian's Wall. In: Roman Shipping and Trade: Britain and the Rhine Provinces. J. du Plat Taylor and H. Cleere (eds.) Research Report No. 24. London: The Council of British Archaeology.

Colls, D. Etienne, R. et al. 1977. L'Epave Port-Vendre II et le commerce de la Bétique à l'époque de Claude. Archaeonautica 1: 23-40.

Costantini, D.A. 1933. Minturnae. Bolletino dell'Associazione Internazionale Studi Mediterranei 1-2: 6-12.

Cuppers, H. 1969. Die Trierer Romerbrucken. Mainz an Rhein: P. von Zabern.

D'Arms, J.H. 1981. Commerce and Social Standing in Ancient Rome. Cambridge, Mass.

de Boe, G. 1978. Roman boats from a small river harbour at Pommeroeul, Belgium. In: Roman Shipping and Trade. J. du Plat Taylor and H. Cleere (eds.) Research Report No. 24. London: The Council of British Archaeology.

de Rossi, G. 1980. Lazio Meridionale. Roma: Newton Compton.

de'Spagnolis, M. 1981. Minturno. Edizioni di Odisseo, Itri, Italy.

Dyson, S.L. 1976. Cosa: the utilitarian pottery. Rome: Memoirs of the American Academy 33.

Feugére, M. 1981. Découvertes au quartier de Villeneuve, Fréjus (Var): le mobilier metallique et la parure. Documents d'Archéologie Méridionale IV: 137-168.

Fiori, P. 1972. Etude de l'Epave A de la Garoupe dites des dolia. Cahiers d'Archéologie Subacquatique I: 37-44.

Flambard, J.M. 1983. Les collèges et les élites locales à l'époque republicaine d'après l'example de Capoue. Les "Bourgeoise" Municipales Italiennes aux IIe et Ier Siècle av. J.C. Centre Jean Berard, Naples: 75-90.

Forbes. R.J. 1964. Studies in Ancient Technology IX. Leiden: E.J. Brill.

Fredericksen, M.W. 1981. I cambiamenti delle strutture agrarie nella tardo repubblica: La Campania in Società romana e produzione schiavistica. In: L'Italia: Insediamenti e Forme Economiche. A. Giardina and A. Schiavone (eds.) Bari: Laterza.

Frier, B. 1969. Points on the topography of Minturnae. Historia XVIII: 510-12.

Fryer, J. 1973. The harbour installations of Roman Britain. In: Marine Archaeology. D.J. Blackman (ed.) London: Butterworth and Co. Ltd.

Gaitzak, W. 1980. Eiserne Romische Werkeuge I and II. Oxford.

Gianfrotta, P.A. 1982. Archeologia sott'acqua. Rinvenimenti sottomarini. Etruria Meridionale, in Bollettino d'Arte suppl. 4: 13-36.

Gianfrotta, P.A. and Pomey, P. 1981. Archeologia Subacquea. Milan: A. Mondadori.

Gibbons, D.J.L. and Parker, A.J. 1986. The Roman wreck of c. A.D. 200 at Plemmirio, near Siracusa, (Sicily): Interim report. The International Journal of Nautical Archaeology 15: 267-304.

Grace, V. and Savvatianou-Petropoulakou, M. 1970. Les timbres amphoriques grecs. Delos vol. xxvii: l'Ilot de la Maison des Comédiens. P. Bruneau, et al. (eds.) Paris: E. Boccard: 277-382.

Grenier, A. 1934. Manuel d'Archéologie Gallo Romaine VI, 2. Paris: A. Picard.

Harnett, E. 1986. The Sculptures of Ancient Minturnae (doctoral dissertation) Bryn Mawr College, Bryn Mawr, Pennsylvania.

Hesnard, A. 1984. Epave du Grand Ribaud D: 0-10 de notre ère. Archéologie Sous-marine. Nantes: Musées du Chateau des Ducs de Bretagne: 79-80.

Johnson, J. 1933. Excavations at Minturnae II. Rome: University Museum at the University of Pennsylvania. 1935. Excavations at Minturnae I. Philadelphia: University Museum at the University of Pennsylvania.

Joncheray, J.P. 1970. L'Epave "C" de la Chrétienne. Gap.

Lachmann, K. 1848. Gromatici Veteres. Berlin: G. Reimer.

Lake, A.K. 1934-35. Campana supellex: the pottery deposit at Minturnae. Bollettino dell'Associazione Internazionale Studi Mediterranei V: 97-136.

Laures, F. 1985. The line or net 'free-er' ring. The International Journal of Nautical Archaeology 14: 80- 82.

L'Hour, M. and Long, L. 1984. L'Epave de l'ilot Barthélémy, de Ier s. av. J.C., etc. Archéologie Sous-Marine, Nantes: 54-57.

Martelli, M. 1982. Archeologia subacquea in Toscana, Bollettino d'Arte suppl. 4: 37-86.

Meneghini, R. 1985. Attività e installazione portuali lungo il Tevere. La Riva dell'Emporium. In: Misurae 5. Modena: 162-71.

Milne, G. 1985. The Port of Roman London. London: B.T. Basford, Ltd.

Mingazzini, P. 1933. Il Santuario della Dea Marica alle Foci del Garigliano. Monumenti Antichi XXXVII: 694-933.

Mocchegiani Carpano, C. 1982. Tevere. Premesse per Una Archeologia Fluviale. Bollettino d'Arte suppl. 4: 150-170.

Oswald, F. and Pryce, T.D. 1920. An Introduction to the Study of Terra Sigillata. London: Longmans, Green and Co.

Owen, D.I. 1971. Excavating a classical shipwreck. Archaeology 24: 123.

Oxé, A. and Comfort, H. 1968. Corpus Vasorum Arretinorum. Bonn: R. Habelt.

Panella, C. 1970. In: Ostia II. Studi Miscellanei 16. Roma.

Panella, C. 1973. In: Ostia III. Studi Miscellanei 21, 1 and 2. Roma.

Pasquinucci, M. and Rossetti. G. 1986. Pisa and Portus Pisanus from Antiquity to the Middle Ages. In: Cities on the Sea - Past and Present. Summaries. Haifa, Israel.

Peacock, D.P.S. 1977. Pottery and Early Commerce. Chacterization and trade in Roman and later ceramics. London: Academic Press. 1982. Pottery in the Roman World. London: Longman.

Pleket, H.W. 1983. Urban elites and business in Trade. In: The Ancient Economy. P. Garnsey, K. Hopkins and C. Whittaker (eds.) Berkeley: University of California Press.

Remmelzwaal, A. 1978. Soil Genesis and Quaternary Landscape Development in the Tyrrhenian Coastal Area of South-Central Italy (doctoral dissertation). University of Amsterdam.

Renfrew, C. 1984. Approaches to Social Archaeology. Cambridge: Harvard University Press.

Rostowzew, M. 1903. Tessararum Urbis Romae et Suburbi Plumbearum: Sylloge. Petersbourg.

Rougé, J. 1981. Ships and Fleets of the Ancient Mediterranean. Middletown, Conn.: Wesleyan University Press.

Ruegg, S.D. 1983. The underwater excavation in the Garigliano River: final report 1982. The Roman port and bridge at Minturnae, Italy. The International Journal of Nautical Archaeology 12: 203-218.

Ruegg, S.D. 1986. Appendix to the line or net free-er ring. IJNA 15: 169.

Ruegg, S.D. The Catalogue of Artifacts from the Garigliano River, Italy (in preparation).

Ruegg, S.D. The Final Report on the Underwater Excavation in the Garigliano River 1966-1981 (in preparation).

Salmon, E.T. 1967. Samnium and the Samnites. Cambridge University Press.

Salmon, E.T. 1970. Roman Colonization Under the Republic. Ithaca, N.Y.: Cornell University Press.

Santamaria, C. 1961. Travaux et découvertes sur l'épave "A" du Cap Dramont. In: Atti del II Congresso Internazionale di Archeologia Sottomarina. Bordighera.

Santamaria, C. 1975. L'Epave A du cap Dramont (Saint Raphaël) fouilles 1971-74. Revue Archeologique de Narbonnaise 8: 185-98.

Schmiedt, G. 1975. Gli Scali Fenicio-Punici I Porti della Magna-Grecia. Firenze: Istituto Geografico Militare.

Scrinari, V. 1979. Le Navi del Porto di Claudio. Roma: Tipografia Centenari.

Sevink, J. Remmelzwaal, A. and Spaargaren, O.C. 1984. The Soils of Southern Lazio and Adjacent Campania. University of Amsterdam.

Shaw, J.W. 1972. Greek and Roman harbourworks. In: A History of Seafaring. G.F. Bass (ed.), New York: Walton and Co.

Siena, S.L. 1977. Appunti su alcuni tipi di anfore Lunensi. In: Collection de l'Ecole Francaise de Rome 32: 207-230.

Staedler, E. 1942. Zu des 29 new aufgefunden inschriftsstelen von Minturno. Hermes 77: 149-196.

Taylor, D.M. 1957. Cosa: Black-Glaze Pottery. Rome: Memoirs of the American Academy in Rome 25.

Tchernia, A. 1971. Les amphores vinaires de Terraconaise et leur exportation au debout de l'Empire. Archivo Espanol de Arqueologia 44: 38-84.

Tchernia, A. 1981. Quelque remarques sur le commerce du vin et les amphores. In: The Seaborne Commerce of Ancient Rome. J. D'Arms (ed.) Cambridge, Mass.

Tchernia, A. 1983. Italian wine in Gaul at the end of the Republic in Trade. In: The Ancient Economy. P. Garnsey, K. Hopkins and C.R. Whittaker (eds.) Berkeley: University of California Press.

Tchernia, A., P. Pomey, A. Hesnard et coll. 1978. L'Epave Romaine de la Madrague de Giens (Var). Gallia, Suppl. XXXIV, Paris.

Tommasino, G. 1942. Aurunci Patres. Gubbio, Italy.

Tylecote, R.F. 1976. A History of Metallurgy. London: The Metals Society.

Ucelli.G. 1940. Le Navi di Nemi. Rome: La Libreria dello Stato.

Vita-Finzi, C. 1969. The Mediterranean Valleys: Geological Changes in Historical Times. Cambridge, England.

Walker, D.S. 1967. A Geography of Italy. London: Methuen & Co. Ltd.

Walters, H.B. 1899. Catalogue of the Bronzes, Greek, Roman and Etruscan in the British Museum. London.

Young, G.F. 1923. The Medici. New York: E.P. Dutton and Co.

SERGE XIMENES, MARTINE MOERMAN

The Roman Harbour of Laurons: Buildings and Structures

ABSTRACT

The Roman site of Laurons includes an interesting choice of harbour structures, such as dykes, piers, wharfs, ware-houses, making it possible to reconstitute approximately one millenium of its activity.

THE SITE

The site of Laurons is situated in the territory of Martigues, on the Mediterranean coast of France (Figs. 1 and 2), in a rich archaeological region which includes Massilia (Marseilles), Aqua sextiae (Aix), Fossis Marianae (Fos-sur-Mer). Abundant and diversified findings in this region indicate intensive maritime and commercial activities from Roman times.

The cove of Laurons comprises three rocky inlets (calanque), two of which are extended by small valleys (Fig. 3). Three hills border on the site. The first one is localized between northern and eastern valleys and the second between eastern and southern creeks. The third hill, behind the southern creek, protects the harbour from prevailing east winds. The northern creek (Fig. 3) was protected by an outer dyke (H) and an inner pier (A). Around this creek cut stone wharfs (Q) were arranged. On the eastern shore, two groups of buildings (B), parallel to the coast, were supplied by an aqueduct bringing water from the Font de maure springs (R). The eastern creek was protected by a pier (C) perpendicular to its northern shore. Behind this protection, a wharf (E) and wooden piles, representing the vestiges of landing stages (F), were found. The southern creek was protected by a dyke (I). A rectangular building lies on its southern shore (G). Eight wrecks have been discovered in the harbour (D,J,K,L,M,N,O,P), most of them situated between the southern dyke (I) and the eastern creek. One wreck was found outside of the southern dyke (S). A dump, extended over approximately 3,000 sq. m. was found inside the harbour, close to the southern dyke.

In 1968, most of the northern creek structures, except for the dyke (H) and one building, were destroyed and the cut stone wharfs were dug up.

Today dykes, piers and wharfs are submerged. The southern creek, as well as its rectangular building, is partially submerged and partially filled in.

HISTORY OF DISCOVERY

In 1953, A. Lotti, owner of the site, carried out the first archaeological drillings around the northern inlet and discovered a large Roman building made up of several construction phases. He also pointed out the existence of an ancient harbour. In 1966, L. Monguilan discovered the Roman dykes by aerial survey. In 1968, the northern creek was threatened by the construction of a steam regenerating station. A. Dumoulin, Assistant Director of Antiquities, carried out a rescue excavation, digging up all the buildings and drawing up the plans of this creek. He also found many fragments of ceramics, glass and metal (Euzennat 1968). The northern creek buildings were then destroyed and abandoned. In 1977, S. Ximenes discovered the first wreck. Seven other wrecks (Gassend, Liou, Ximenes 1984; Ximenes, Moerman 1987) were then found and studied by GRASM(1). The excavations of the dump were started in 1982, and those of harbour structures in 1983. These excavations are still being carried out.

ANCIENT AND PRESENT TERRESTRIAL EXCAVATIONS

Northern creek. A. Lotti(2) discovered the buildings (B) and noticed the presence, in the west of these structures, of a buried section of the ancient harbour (Fig. 4) Aerial photographs (Fig. 18) and plans showed the layout of the dock and Lotti's drill-holes. One of these photographs (Fig. 19) showed one cut stone wharf. This wharf probably bordered the western shore of the dock. It was at least two cut stones high and two cut stones wide. According to the scale of the photograph, cut stones were approximately 30 cm high, 55 cm long and 25 cm wide(3). Mortar jointing can be seen. Furthermore, plans showed an irregular quadrangular construction on the western shore. Lotti hypothesized that this structure could have been the foundations of a Roman tower (T).

The first building of the northern creek was made up of several rooms around two groups of hypocaustes (Fig. 20). A third part could have been a storehouse. Behind each of the two groups of rooms was a courtyard and a well. It seems that the two hypocaustes worked jointly. The building was drained by several sewers. Different ages of construction were found in this building: the earliest construction occurred during the 1st century B.C.E., reorganization during the period of Augustus and, extensions during the Lower Empire. In the second building a room with an apse can be seen.

Southern creek. (Fig. 5). The big rectangular building, situated on the present shore, was separated into two unequal parts by a longitudinal wall. The construction was made over a drainpipe system (fig. 21). The role of this building is unknown. According to construction type and some ceramics found during the excavation, the building may have been constructed at the beginning of the 1st century C.E. The submerged part of the building is poorly preserved.

UNDERWATER EXCAVATIONS

Dykes (Fig. 3). The northern dyke, the larger of the two, was the more exposed to waves. It is 104 m long. The first 20 metres functioned as a breakwater. The dyke is 15 m wide. Its visible height is 2.50 m at the southern extremity and 1 m in the northern part. Current drill holes will determine the real height of this construction. This dyke, made of large cut blocks without mortar, has stayed together throughout the centuries.

The southern dyke, 75 m long, was built with uncut stones and has fallen apart. Its present dimensions are 24 m wide in the middle part and 2.50 high at the northern extremity. Initial dimensions may be estimated at 15 m for width and 2.50 m for height. In order to avoid the collapse of the construction, slabs have been vertically and horizontally placed at the northern end of the dyke, as is indicated in Figs. 10 and 11.

Both northern and southern dykes protected the harbour from north and west winds: the Mistral and the the Largade.

Piers. The pier built on the west shore(4) of the northern creek (Fig. 23) was made of two parallel lines of blocks. Its length was 13.5 m and its width 2 m. The number of rows of blocks has not been determined. As shown in Fig. 13, these cut blocks were 100 x 65 x 20 cm or 100 x 65 x 25 cm. Other cut blocks have been found around the pier. Their sizes were identical or smaller (75 x 50 x 20 cm).

The eastern creek pier (Fig. 24), perpendicular to the northern shore is being excavated. Its length is 19 m. Its width varies from 9.50 m near the shore to 7 m at its southern extremity. Its height varied from 1.80 to 0.75 m, and was made of several rows of cut stones without bonds.

Piles. Thirty wooden piles have been found. Most of them were situated in the extension of the east creek pier (Figs. 8 and 25) suggesting the existence of a landing stage. Twenty-four piles have circular sections. Their diameters vary from 10.5 to 20 cm(5). The six other piles are rectangular. Their sizes vary from 9 to 14 x 13.5 to 18 cm.

Cut stones (Figs. 8 and 9) In the east creek, between the northern shore and the wharf, 120 cut blocks were found. Seven different types of blocks have been numbered, some of them constituting a visible alignment. The role of these blocks is not clearly known.

The wharf. The east creek wharf is situated along the north shore (Fig. 8) Its orientation is east-west. It is composed of two parts: a wooden caisson which was used for wharf building and a cut stone construction.

The wooden caisson is 22.90 m long and 2.20 wide. It is composed of several elements. The bottom was a floor made up of wide fitted planks. These planks, parallel to the small sides of the wharf, are nailed under a horizontal frame(6) composed of beams (Fig. 10). The upper side of the beams presented deep grooves. Within the frame two series of horizontal beams strengthened the floor. Thirty two piles, situated all around the caisson, ensured its rigidity. These rectangular piles ended in dovetails cut in their outer parts(7). These dovetails fitted into the frame beams and were nailed to them. The inner parts of the piles were positioned on these beams. The piles were marked by Roman numerals with triangles (Figs. 11, 26 and 27) on their lateral sides. Corresponding numerals found on the beams, near the dovetails, suggest that these marks were used during caisson assembly. On the north side of the wharf, piles, regularly numbered from II to X, were found. The lateral walls of the caisson were made with stakes vertically fitted into the grooves of the frame beams. The grooves were broken off before and after pile insertion. Thus, the stakes situated close to the piles were partially positioned on the beam (Fig. 12). Groups of cylindrical piles were vertically planted around the wharf. Their role was probably to strengthen the junctions between the frame beams during the construction. Two types of junctions were used. The first one is a mortise-and-tenon angle junction (Fig. 13). The second one assembles long side beams: the lower part of one beam is positioned under the upper part of the other (Figs. 14 and 28) with a complex imbrication ("Trait de Jupiter").

Hypothesis for the building of the caisson: the caisson was built out of the water. When the frame beams were assembled, the grooves were situated on the ground. Then, the floor was nailed and the construction turned around. The piles and stakes were then set in place (Fig. 15). One can hypothesize that the structure was maintained by an upper frame similar to the lower one (Fig. 16). Then, the caisson was probably brought to water with ropes. Two ropes were found below the wharf during the excavation. During the construction of the wharf, the caisson progressively became submerged because of the weight of the cut stones.

Present state of the wood caisson: the beam frame and the floor are well preserved as are the northern piles. Only one stake was found on the northern face.

Stone construction: the stone construction was made of little cut stones and mortar. Western and southern sides, exposed toward the sea, were sloping and covered by mortar. Other sides were vertical and apparently had stones (Figs. 29 and 30).

V - VARIATION OF SEA LEVEL IN THE SITE OF LAURONS

In addition to the general rise of the Mediterranean Sea level since Roman times, the Fos gulf, in which the cove of Laurons is situated, is characterized by a localized and particularly important submergence. This submergence, estimated as approximately 2 m, may be attributed to the subsidence of recent sedimentary strata in ancient crevices. The submergence has had important effects on the Roman harbour. As shown in Fig. 17, the dykes disappeared into the sea at an angle. The shore retreated to a distance of 120 m on the northern dyke and 30 to 35 m on the southern dyke (Fig. 3). The wharf was submerged to a depth of approximately 1 m and remained in horizontal position. The shore close to the wharf retreated by 3 to 5 m(8).

Dyke submergence may have been responsible for a progressive infilling of the harbour. According to archaeological material discovered in the dump, the last signs of harbour activity may date from the 7th century C.E.

VI - HARBOUR ACTIVITY, DATINGS AND IDENTIFICATION

The harbour activity was diversified. Laurons cove was probably a stopping place for boats sailing along the coast and coming from Italy, Spain or, more modestly, from Marseilles or Arles. The Laurons harbour also had commercial functions. Archaeological materials coming from Gaul, Spain, Italy, North Africa, the Greek islands and Alexandria, were found in this site. The third activity of this harbour was the transport of cut stones coming from quarries situated a few kilometres northwards (Fig. 31). Ruts, found in the hill, show the way by which the cut stones were brought to the Laurons. Two cut stone wrecks have been discovered in the site. The activity of Laurons harbour was probably important. Indeed, its surface area, estimated to more than 100,000 sq. m., was comparable to that of the harbour of Forum Julii (Frejus) or Centumcellae (Scranton, Shaw and Ibrahim 1978, p. 14), much better known in ancient maritime history. According to the materials found in the dump, the Laurons harbour activity may have lasted from the middle of the 3rd century B.C.E. to the end of the 7th century C.E. The Lower Empire may have been the period of its apogee. Antonin's Maritime Itinerary points out several harbours in this area, some of which have been identified, such as Incarrus Positio (Carry-le-Rouet) and Fossis Marianis. Between these two harbours another name is mentioned: Dilis Positio. R. Lugand (Lugand 1926) defined "positio" as a natural anchorage without built structures, as distinct from a harbour. One could hypothesize that the cove of Laurons was called Dilis Positio before the Augustus period and kept this name until the 1st century C.E., the period in which Antonin wrote the Itinerary. Unfortunately, no epigraphic clue has been found in the Laurons.

NOTES

1. Groupe de Recherche Archéologique Sous-Marine, Clinique de la Résidence du Parc, Rue Gaston Berger 13010 Marseille. Comité de Provence de la Fédération Française d'Etudes et de Sports sous-marins.

2. We thank Mr. A. Lotti and Mr. Monguilan for the information provided and for their photographs.

3. These measures are approximate.

4. Excavated in 1968 by A. Dumoulin and later destroyed.

5. These measures are only valid for the visible part of the piles.

6. Section: 14 x 18 cm for the short sides and 11 x 18 for the longer sides.

7. Section: 13 x 11 cm.

8. This distance is calculated without taking into account modern arrangements.

REFERENCES

Bellet, M. 1979. Guide Archéologique des Rives de l'Etang de Berre. Coll. La Méridienne. Aix-en-Provence, Edisud, 68-69.

Benoît, F. 1954. Informations Archéologiques: XII ème circonscription, Martigues. Gallia XII: 433-434.

Benoît, F. 1962. Informations Archéologiques: Circonscription, Aix (Région Sud): Martigues. Gallia XX: 686-688 and 690.

Euzennat, M. 1969. Informations Archéologiques, Circonscription de Provence (Région Sud), Martigues. Gallia XXVII: 430-432.

Gassend, J.M., Liou, B., Ximenes, S. 1984. L'épave 2 de l'anse des Laurons. Archaeonautica 4: 75-105.

Lugand, R. 1926. Note sur l'itinéraire maritime de Rome à Arles. MEFR 46: 124-139.

Office Municipal Socio-Culturel. 1977. Documents de l'Histoire de Martigues de la Préhistoire au XIX ème Siècle. Martigues. Im. Maury, 35-37.

Scranton, R., Shaw, J.W. and Ibrahim, L. 1978. Kenchrai, Eastern Port of Corinth. Vol. 1 Topography and Architecture. Leiden, E.J. Brill.

Ximenes, S., Moerman, M. 1985. Le port gallo-romain de l'Anse des Laurons Martigues (Bouches-du-Rhône). In: 20 Ans d'Archéologie Sous-Marine sur les côtes de France. DRASM (Ed.) Nantes and Perros-Guirrec.

Ximenes, S., Moerman, M. and Rapsaet, G. 1985. Archéologie sous-marine en provence, le port romain de l'Anse des laurons (Bouches-du-Rhône). Annales d'Histoire de l'Art et d'Archéologie (ULB) VII. 135-145.

Ximenes, S. and Moerman, M. 1986. Dans l'Anse des Laurons: un port romain et sept épaves. Etudes et Sports Sous-Marins 85: 24-25.

Ximenes, S. and Moerman, M. 1987. Les épaves I, III et IV du port romain de l'Anse des Laurons. Cahiers d'Archéologie Sub-Aquatique VI: 195-206.

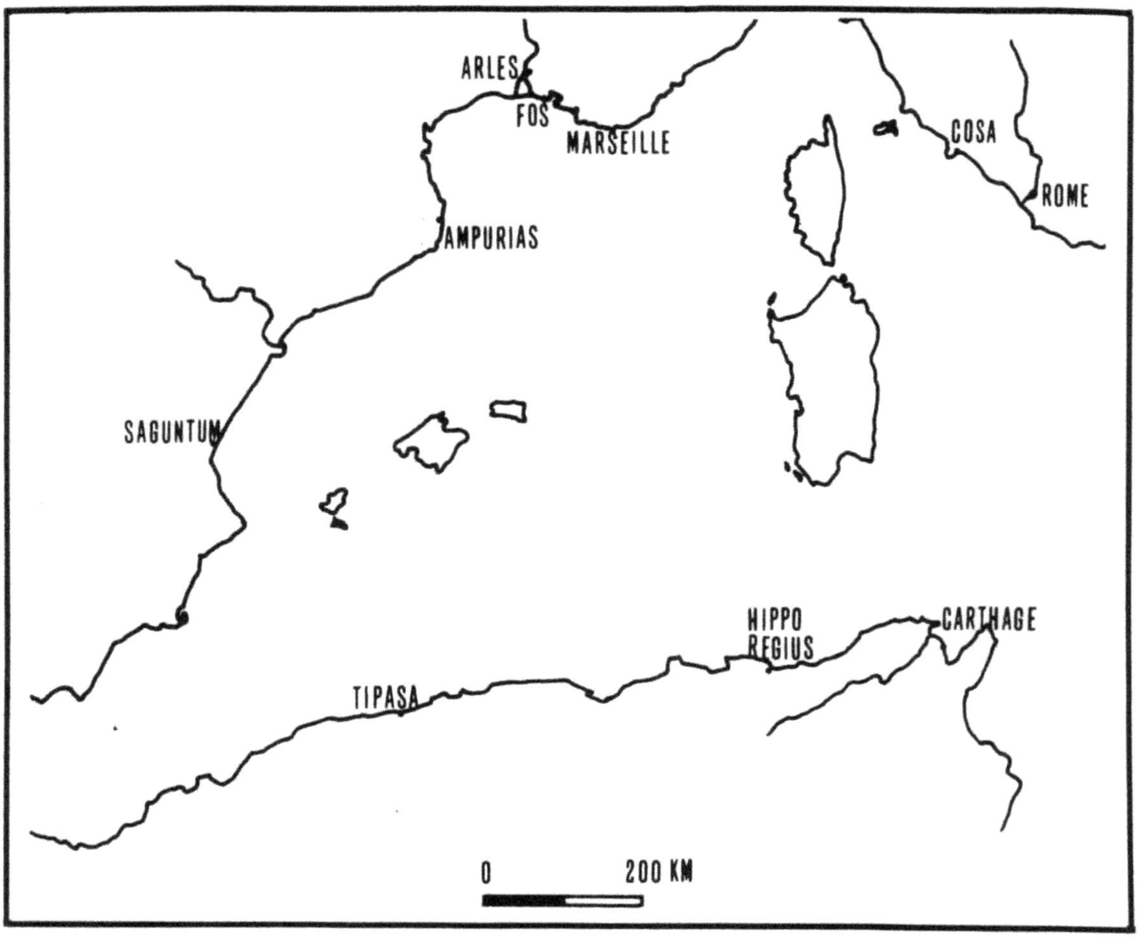

Figure 1: Western Mediterranean sea (S.X. - M.M.)

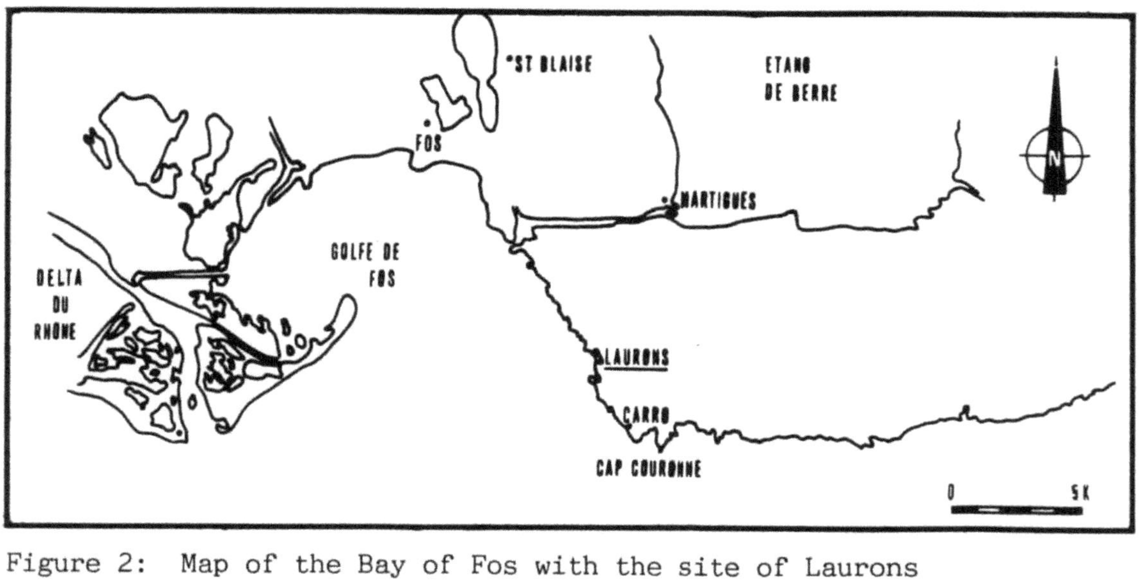

Figure 2: Map of the Bay of Fos with the site of Laurons
(S.X. - M.M. 1986)

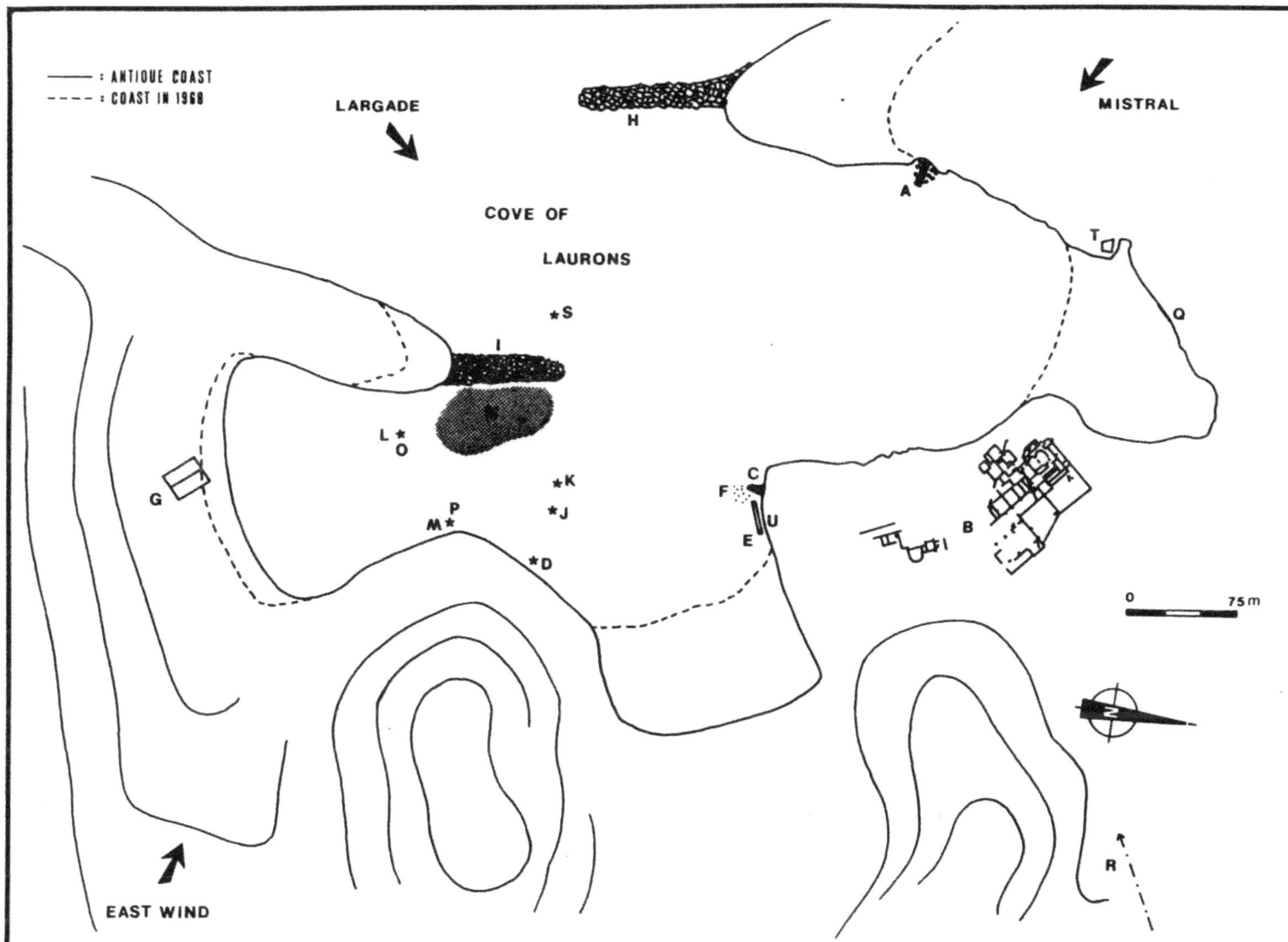

Figure 3: The site of Laurons (S.X. - M.M. 1987)

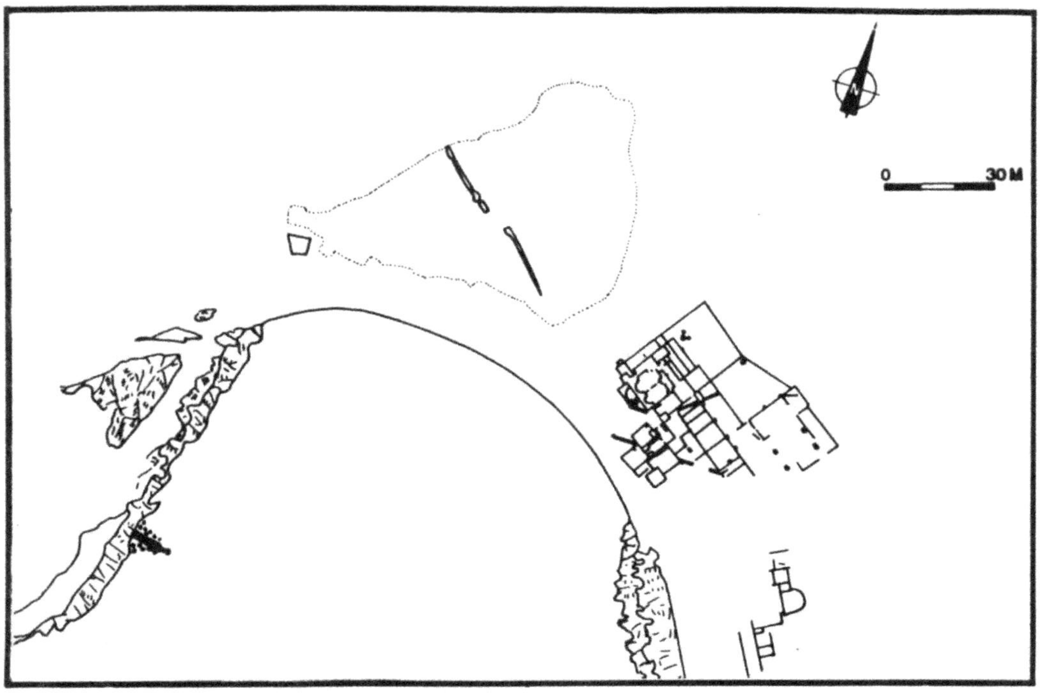

Figure 4: The northern creek (EDF - Dumoulin 1968).

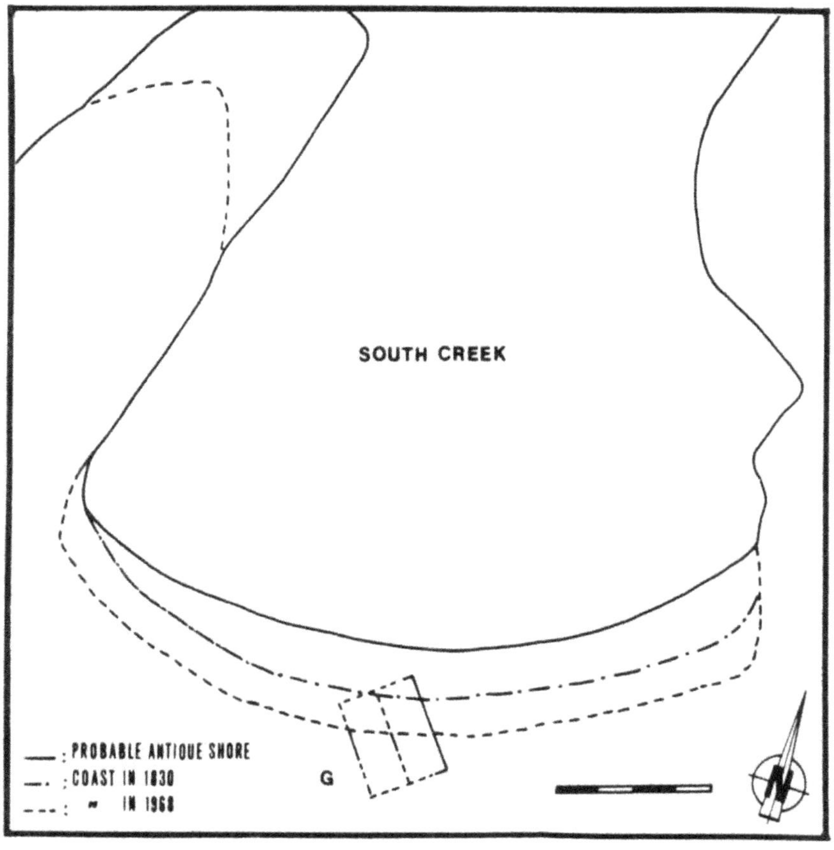

Figure 5: The southern creek (S.X. - M.M. 1987)

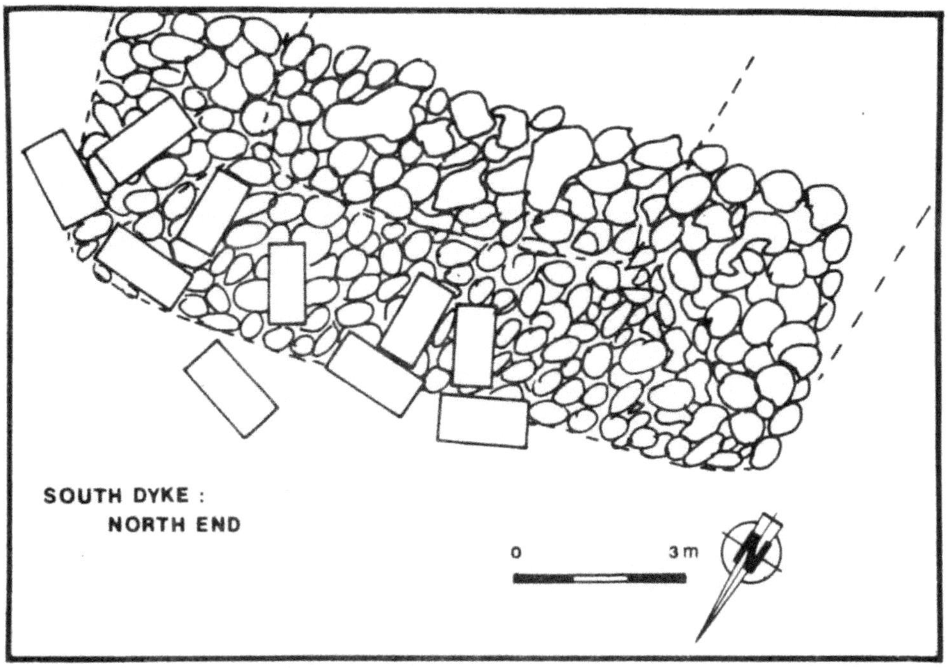

Figure 6: Slabs on the northern end of the southern dyke
(S.X. - M.M. 1985)

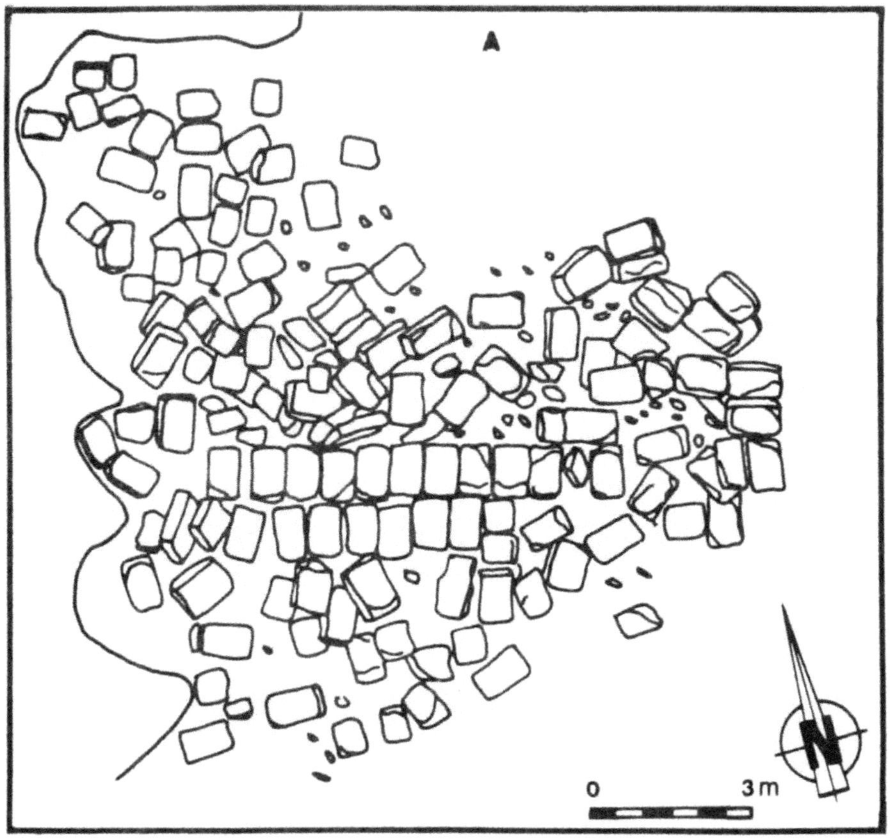

Figure 7: The pier of the northern creek (M. Rival 1968).

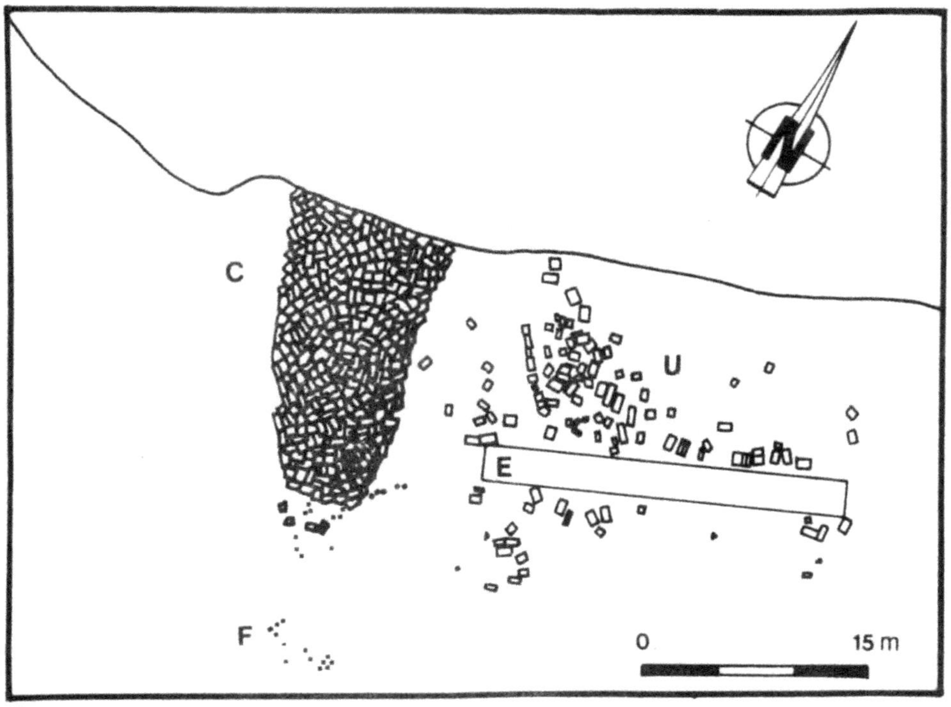

Figure 8: Landing stage of the eastern creek
(S.X. - M.M. 1986)

Figure 9: Blocks around the wharf in the eastern creek
(S.X. - M.M. 1986)

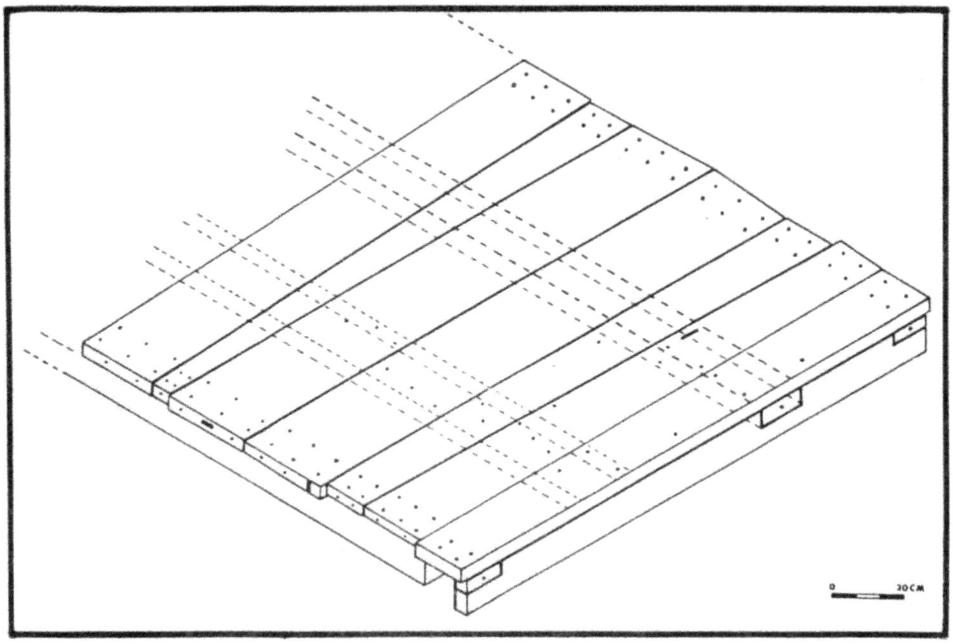

Figure 10: Caisson nails on longitudinal and frame beams.
The figure also shows a stage of the construction of the caisson. (S.X. - M.M. 1986)

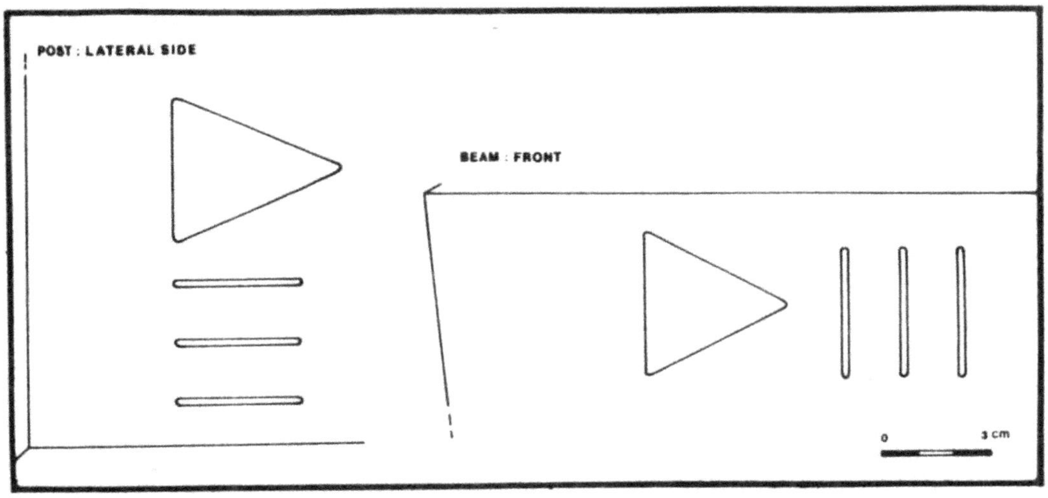

Figure 11: Roman numbers on a pile and on a beam (S.X. - M.M. 1986)

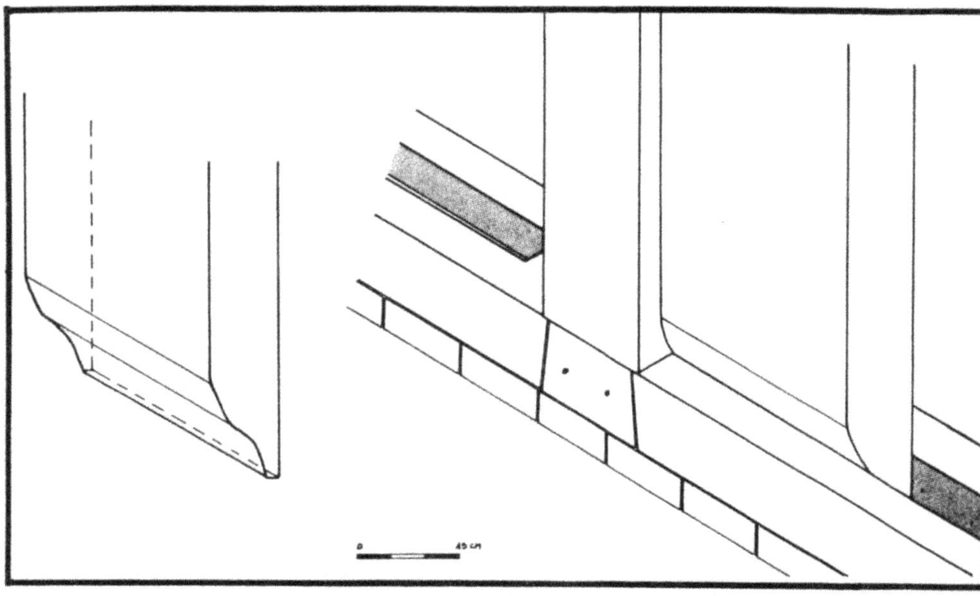

Figure 12: Stake isolated and placed into the groove against a pile. (S.X. - M.M. 1986)

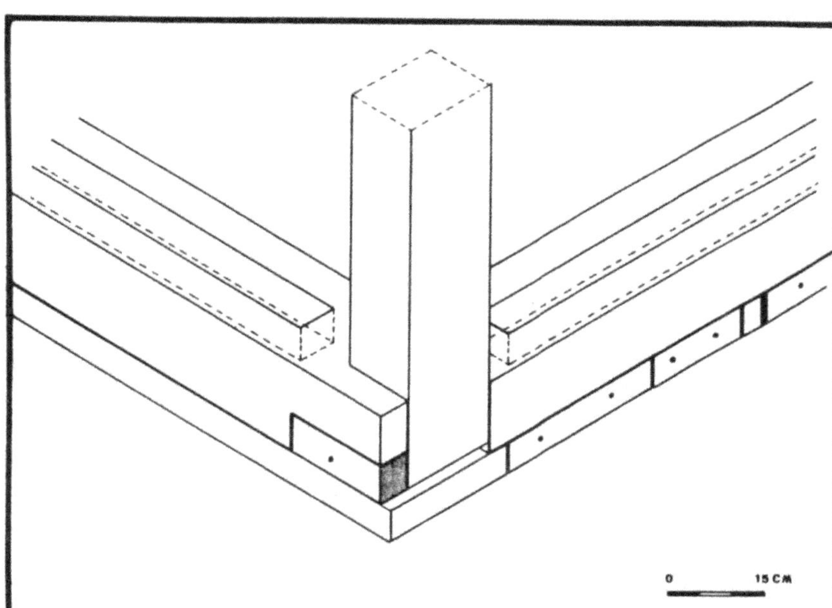

Figure 13: Angle junction of the caisson with short side (left) and long side (right). The pile is set into the junction. (S.X. - M.M. 1985)

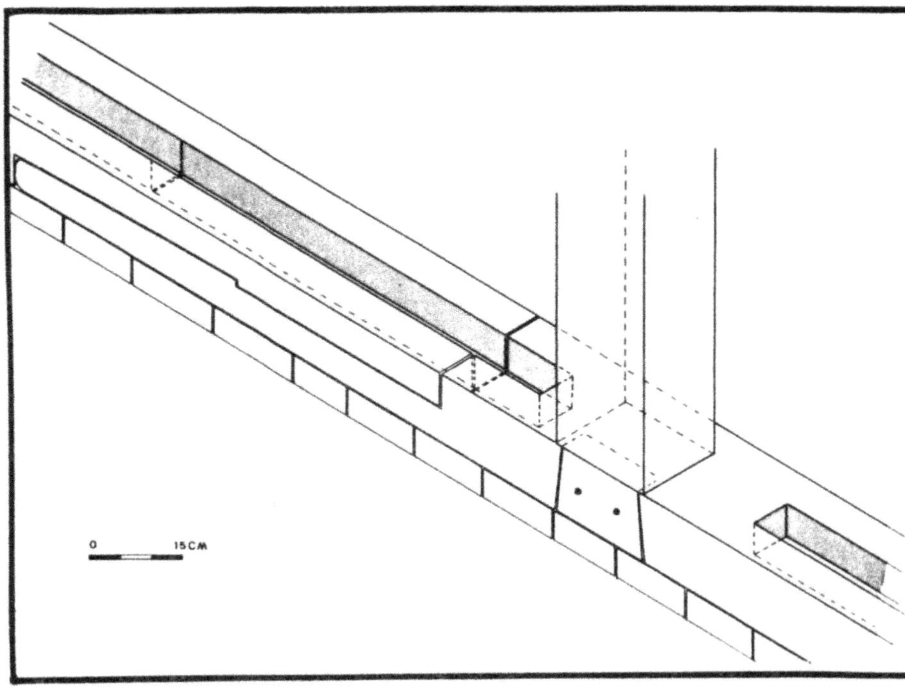

Figure 14: Junction of long side beams with the ends of the floor planks and the pile. (S.X. - M.M. 1986)

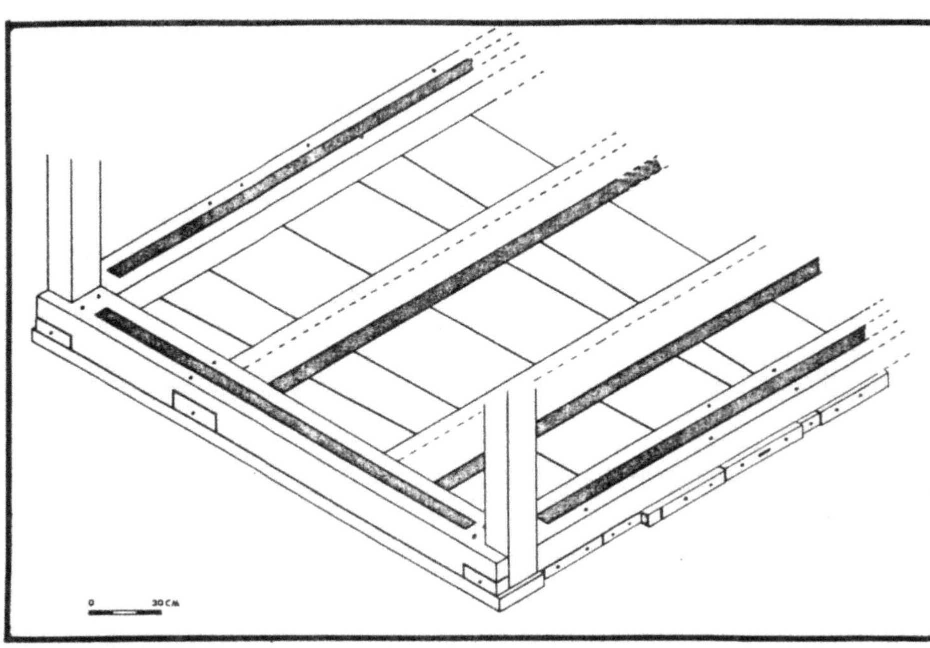

Figure 15: Western end of the caisson with the floor, the beam frame and two piles. (S.X. - M.M. 1986)

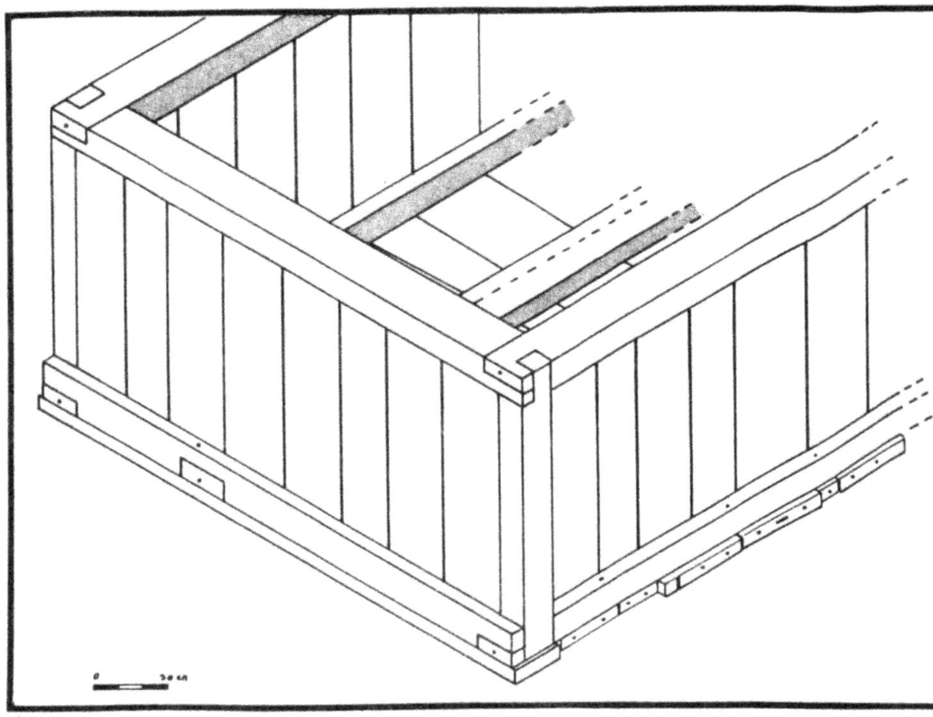

Figure 16: Hypothetical reconstitution of the western end of the caisson. (S.X. - M.M. 1986)

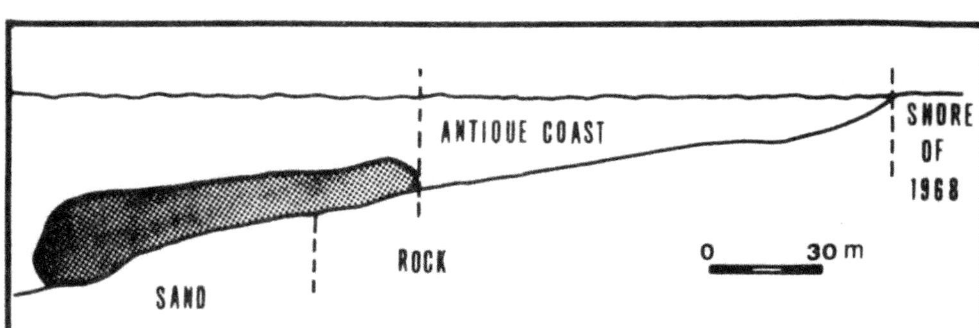

Figure 17: Section of the northern dyke and limits of the ancient and 1968 coasts. (S.X. - M.M. 1986)

Figure 18: The northern creek (L. Monguilan).

Figure 19: Northern creek: wharf (A. Lotti - 1954).

Figure 20: Northern creek - Excavations of A. Dumoulin 1968.

Figure 21: Excavations of the "G" building: walls and drain-pipe (S.X. - M.M. 1986).

Figure 22: Vertical slabs on the northern end of the southern dyke (S.X. 1985).

Figure 23: The pier of the northern creek (A. Dumoulin 1968)

Figure 24: Pier of the eastern creek (A. Lotti 1954).

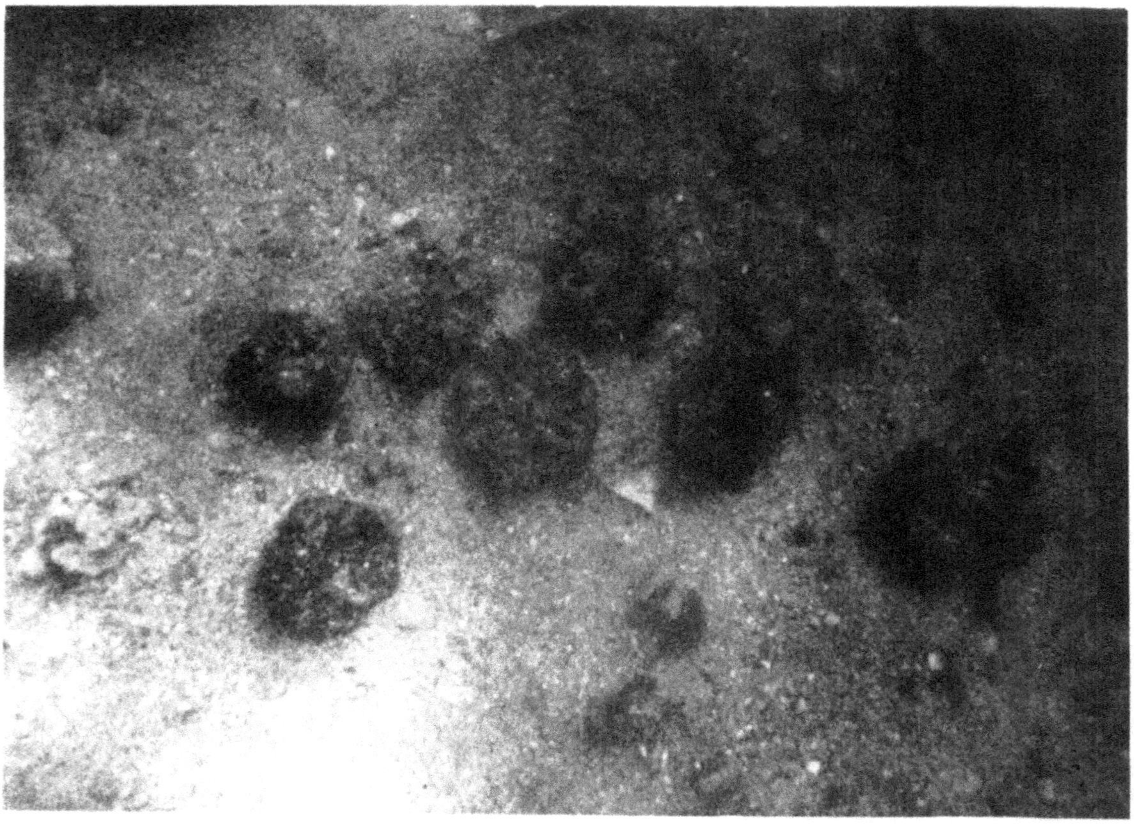

Figure 25: Cylindrical piles of the eastern creek (S.X. 1985)

Figure 26: Roman number on a pile of the northern side.
(M. Debatty 1986).

Figure 27: Roman number on a beam of the southern side.
(S.X. 1985)

Figure 28: The junction of two beams of the southern long side. (S.X. 1985)

Figure 29: Stone construction and caisson in the western end. (S.X. 1985)

Figure 30: The northern side (M. Debatty 1986).

Figure 31: Ancient quarries of "Les Olives" nearby the site of Laurons (M.M. 1986).

www.ingramcontent.com/pod-product-compliance
Ingram Content Group UK Ltd.
Pitfield, Milton Keynes, MK11 3LW, UK
UKHW061847041026
15591UKWH00015B/945

9 780860 545194